T0332941

# A SYSTOLIC ARRAY PARALLELIZING COMPILER

THE KLUWER INTERNATIONAL SERIES IN
ENGINEERING AND COMPUTER SCIENCE

HIGH PERFORMANCE COMPUTING
Systems, Networks and Alrogrithms

Consulting Editor

H.T. Kung

# A SYSTOLIC ARRAY PARALLELIZING COMPILER

by

**Ping-Sheng Tseng**
Bell Communications Research, Inc.

with a foreword by

**H.T. Kung**

**KLUWER ACADEMIC PUBLISHERS**
**Boston/Dordrecht/London**

**Distributors for North America:**
Kluwer Academic Publishers
101 Philip Drive
Assinippi Park
Norwell, Massachusetts 02061 USA

**Distributors for all other countries:**
Kluwer Academic Publishers Group
Distribution Centre
Post Office Box 322
3300 AH Dordrecht, THE NETHERLANDS

**Library of Congress Cataloging-in-Publication Data**

Tseng, Ping-Sheng, 1959–
    A systolic array parallelizing compiler / by Ping-Sheng Tseng.
      p.   cm. — (The Kluwer international series in engineering and
computer science ; #106. High performance computing)
    Revision of the author's thesis (Ph.D.)—Carnegie Mellon
University.
    Includes bibliographical references and index.
    ISBN 0-7923-9122-5
    1. Parallel processing (Electronic computers)  2. Compilers
(Computer programs)  I. Title.   II. Series: Kluwer international
series in engineering and computer science ; SECS 106.   III. Series:
Kluwer international series in engineering and computer science.
High performance computing.
QA76.58.T74   1990
005.4′53—dc20                                             90–4751
                                                          CIP

# Contents

# List of Figures

# Foreword

Widespread use of parallel processing will become a reality only if the process of porting applications to parallel computers can be largely automated. Usually it is straightforward for a user to determine how an application can be mapped onto a parallel machine; however, the actual development of parallel code, if done by hand, is typically difficult and time consuming. Parallelizing compilers, which can generate parallel code automatically, are therefore a key technology for parallel processing.

In this book, Ping-Sheng Tseng describes a parallelizing compiler for systolic arrays, called AL. Although parallelizing compilers are quite common for shared-memory parallel machines, the AL compiler is one of the first working parallelizing compilers for distributed-memory machines, of which systolic arrays are a special case. The AL compiler takes advantage of the fine grain and high bandwidth interprocessor communication capabilities in a systolic architecture to generate efficient parallel code.

While capable of handling an important class of applications, AL is not intended to be a general-purpose parallelizing compiler. Instead, AL is designed to be effective for a special class of computations that use arrays and loops. AL relies on the fact that for these computations, the user can easily provide "hints" or mapping strategies to guide the compiler to distribute data structures and loop iterations. Using these hints, the AL compiler generates the local program for each processor, manages the interprocessor communication, and parallelizes the loop execution. A fundamental contribution of AL, which goes beyond its current implementation, is the identification of what to capture in these hints that the user can easily provide and the compiler can effectively use.

AL has proven to be extremely effective in programming the Warp systolic array developed by Carnegie Mellon. AL was used to port the nontrivial LINPACK QR (SQRDC), SVD (SSVDC), LU (SGEFA), and back substitution (SGESL) routines to Warp all in one person-week. AL has been used in several applications, including the porting of a large Navy signal processing application to Warp. The AL compiler has also been used to generate parallel code for automatic schedulers that map a large set of high-level signal processing tasks onto Warp.

Carnegie Mellon's experience in programming Warp clearly indicates the effectiveness of special-purpose parallelizing compilers such as AL. These tools, which we also call parallel program generators, can improve a programmer's productivity by several orders of magnitude. In fact, some applications would never have been brought up on Warp if the user did not have access to AL; for these applications explicit programming of interprocessor communication is simply too difficult to be done by a human being. Besides AL, the Warp project has developed several other parallel program generators. One of them, called Apply, has been extensively used to generate parallel code for image processing applications. Both AL and Apply generally produce code better than or as good as hand-written code.

The success of parallel program generators such as AL is encouraging. With them we are assured that programming parallel machines can be as easy as programming sequential machines for some important application areas. In this sense, these programming tools have helped legitimatize our effort in building parallel computers. Besides providing useful tools for programming parallel machines for special applications, these parallel program generators are also significant in giving insights into information that more general-purpose parallelizing compilers of the future need to capture.

The book is an outgrowth of Ping-Sheng Tseng's Ph.D thesis from Carnegie Mellon. The book gives a complete treatment of the design, implementation and evaluation of AL. I am pleased to write the Foreword for this outstanding piece of work and hope that the book will inspire researchers to further this very important area of parallel processing.

*H. T. Kung*
April, 1990
Pittsburgh, PA

# Acknowledgements

This work evolved from my thesis research at Carnegie Mellon University. I want to thank my thesis advisor H. T. Kung. When I started this research project, Kung was the only one who believed that I could make this thesis happen. Without his vision, support and encouragement, I might have changed my thesis topic before I started. Thanks to Monica Lam and Peter Lieu. Monica implemented the W2 compiler directives for me. Peter has been very responsive to my W2 bug reports. Without their help, all the experimental work might have never been completed. The members of my thesis committee are H. T. Kung, Gary Koob, Thomas Gross, and George Cox. Their invaluable comments helped me to improve the quality of this presentation. I want to thank my former officemate Robert Cohn. Robert read my manuscript several times and helped me to make this presentation much more readable. I would also like to thank the Defense Advanced Research Projects Agency for supporting this work. Finally, I would like to thank my wife Ly-June

and my father Po and my mother Sou-Sou for their support in my
graduate study.

# A SYSTOLIC ARRAY PARALLELIZING COMPILER

# Chapter 1

# Introduction

Programmable systolic arrays are attractive for high speed scientific computing because of their truly scalable architecture. However, systolic array programming has been an intellectually challenging problem since the architecture was made popular by H. T. Kung in the late 1970s. For many years, researchers have been studying compiler techniques for mapping scientific programs to systolic arrays [5, 7, 12, 14]. However, these efforts were focused on hard-wired systolic arrays, and thus many of their results are not directly applicable in the context of programmable systolic arrays. In a hard-wired systolic array, the functions of the processing cells, the size of the array, and the intercell connections are selected to perform a dedicated computation on a fixed problem size. Whereas in a programmable systolic array, the size of the array and the intercell connections are fixed, and the processing cells can be programmed to perform different computations on different problem sizes.

In 1985, a Carnegie Mellon University group led by H. T. Kung built a programmable systolic array, the Warp machine. Since then, the Warp group has been engaged in the real life problems of systolic array programming. At the beginning, the research efforts were focused on the W2 compiler. As Lam [11] wrote

> Compiling processor-oblivious programs to efficient (parallel) code appears intractable at present. In order that efficiency and generality can be achieved, we choose to expose the configuration of the array to the user, while hiding the internal pipelining and parallelism in each cell in the array.

Although the W2 compiler significantly advanced the art of systolic array programming, the user still has to program each cell in the systolic array individually and manage the intercell communication explicitly. The difficulties in manually mapping scientific programs into W2 programs discouraged researchers from using Warp for serious scientific computing applications.

In December 1987, we started tackling the problem of compiling array-oblivious programs into efficient systolic array programs [17]. For this research, we devised a programming language named AL [18] and implemented an AL compiler for Warp. In April 1988, we compiled the LINPACK LU decomposition routine (SGEFA) and achieved a nearly 8-fold speedup for matrices of size 180x180. Within two weeks, we successfully compiled the LINPACK SVD (SSVDC) routine, for which no one had ever successfully written a W2 program by hand. Since then, the AL compiler has been used to generate many scientific computing applications for the Warp machine.

This monograph describes the AL compiler and its applications in the context of the Warp systolic array. In Chapter 2, we promote the idea of a systolic array parallelizing compiler by showing how a systolic array is programmed with and without such a compiler. In Chapter 3, we introduce the notion of data relations, which is the key idea of the AL compiler. In Chapter 4, we describe loop distribution techniques and analyze distributed loop parallelism. In Chapter 5, we discuss implementation details of the AL compiler. In Chapter 6, we evaluate the AL compiler by measuring and analyzing the achieved performance for a set of scientific computing programs. In Chapter 7, we present the conclusion that parallelizing compilers are powerful programming tools for systolic arrays.

# Chapter 2

# Systolic array programming

To appreciate a systolic array parallelizing compiler, we have to show how a systolic array is programmed with and without such a compiler. We begin by presenting the Warp architecture, a linear systolic array of ten powerful processing cells. We then describe the W2 cell programming language and the AL array programming language. W2 is a high level cell programming language which hides the cell architecture from the user, but exposes the array architecture. In W2, the user programs each cell individually and manages intercell communication explicitly. AL is a high level programming language which hides the array architecture from the user. In AL, the user programs the entire systolic array as if it were a sequential computer and the compiler generates parallel W2 programs with intercell communication.

## 2.1   The Warp machine

The Warp machine [2] is a linear systolic array of 10 powerful processing cells as shown in Figure 2.1. The architectural strength of the Warp array comes from its systolic communication channels, the X and Y channels. Through the communication channels, a Warp cell can transfer 2 words to and from its neighboring cells every clock cycle (200ns). This performance translates to a bandwidth of 20 MB/sec for a channel or 80 MB/sec for a cell.

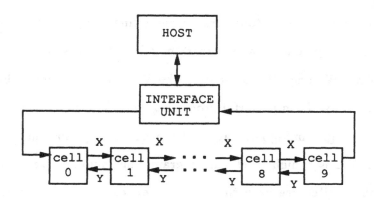

Figure 2.1: The Warp systolic array

Each Warp cell contains one floating-point multiplier, one floating-point adder, one integer ALU and a data memory of 32K words

(128K bytes) as shown in Figure 2.2. All of these components are interconnected through a crossbar switch and controlled by a long instruction word (horizontal microcode). The floating-point units can deliver up to 5 MFLOPS (single precision only) each. This performance translates to a peak processing rate of 10 MFLOPS per cell or 100 MFLOPS for the 10-cell machine. Warp is a MIMD array, each cell has its own local program memory of 8K instruction words (272 bits wide) and its own program sequencer.

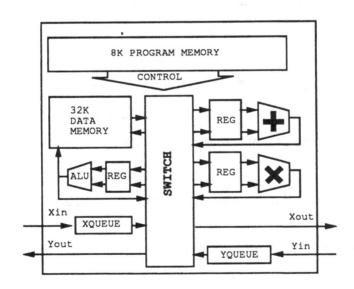

Figure 2.2: The Warp cell architecture

W1 is the assembly language of the Warp cell. A W1 statement is a line of microcode, where each component in the Warp cell is controlled by a dedicated instruction field. In a single instruction cycle, the Warp cell can initiate two floating-point operations, two integer operations, read and write one word in the data memory, read and write 2 words in each register file, receive and send two words from and to its two neighbors, configure the switch to transfer 6 words from the 6 input ports to the 8 output ports, and conditionally branch to a new program location. W1 programming is further complicated by the latency and pipelining of the floating-point units. The result of a floating-point operation will not be delivered until 5 cycles after the operation is issued. In order to secure the result for later use, we have to configure the switch 6 cycles later to route the result and issue a store operations 7 cycles later to store the result. Obviously, W1 is not a language for application programmers. It is a job of compilers to generate W1 code from high level languages.

## 2.2   The W2 programming language

W2 is a high level cell programming language. To the user, W2 is just a simple sequential programming language which has data objects of scalars and arrays, and control constructs of IF-THEN-ELSE, DO-loop, and WHILE-loop. The W2 compiler is a complicated piece of software which uses sophisticated data dependence analysis and loop scheduling techniques [13] to generate high quality W1 code.

The W2 language does not hide the array architecture from the user. The user has to program each Warp cell individually and manage the intercell communication explicitly.

W2 has two communication primitives, SEND and RECV. A SEND statement:

```
SEND(neighbor, word)
```

instructs the cell to send a word to its left neighbor (L) or right neighbor (R). Correspondingly, a RECV statement:

```
RECV(neighbor, word)
```

instructs a cell to receive a word from its left neighbor (L) or right neighbor (R). Synchronization is implicitly done with communication: the SEND operation blocks if the queue (512 words) on the channel is full; the RECV operation blocks if the queue on the channel is empty.

Although W2 supports heterogeneous programming where each cell runs a different W2 program, many Warp applications can employ homogeneous programming where all the cells run the same W2 program and use the cell number to distinguish oneself from the others. For example, the following W2 program broadcasts an array of integers from cell $k$ to all the other cells. In this program, cell $k$ sends out the data bidirectionally to its two neighbors. The cells to the left of cell $k$ receive data from their right neighbors and forward the data to their left neighbors. The cells to the right of cell $k$ receive data from their left neighbors and forward the data to their right neighbors.

```
int b[100];
if (cellid = k) then {
  for j := 0 to 99 do
    {SEND(R, b[j]); SEND(L, b[j]); }
}else{
  if (cellid < k) then {
    for j := 0 to 99 do
      { RECV(R, b[j]); SEND(L, b[j]);}
  }else{
    for j := 0 to 99 do
      { RECV(L, b[j]); SEND(R, b[j]); }
  };
}
```

Notice that the SEND and RECV statements in a W2 program are not operating system calls, they are primitive operators just like the addition and the multiplication operators. Each SEND (or RECV) statement will be translated into a micro-operation and be scheduled into an instruction. There is neither operating system overhead nor message routing overhead in these operations. As a result, it only takes about 110 instruction cycles to execute the above program on a 10-cell Warp machine. Although the Warp architecture supports communication that is both fine-grain (a word a time) and local (neighbor to neighbor), it also performs well on regular global communication such as broadcasting a vector and adding up distributed vectors.

For a more complete example, consider a W2 program for parallel matrix multiplication. One way to do parallel matrix multiplication $C = A \times B$ on Warp is to evenly distribute rows of all the matrices among 10 cells and repeatedly broadcast a row of the matrix B ($brow$) to perform $C = C + A \times [brow]$ ([$brow$] denotes the row submatrix of the matrix B) until the matrix A is multiplied by the entire matrix B. Figure 2.3 shows a W2 program written for this algorithm.

Using W2, we have to program the systolic array by programming its cells. Therefore, we have to manage all the details of program mapping: program decomposition, data distribution, intercell communication, process synchronization, and load balancing. It would be much easier to write the matrix multiplication program directly for the entire systolic array without seeing the cells, then use a compiler to generate W2 programs and manage all the details of program mapping.

## 2.3  The AL programming language

AL is a high level programming language which hides the array architecture from the user. In AL, the user programs the entire systolic array as if it were a sequential computer and the parallelizing

```
float a[10][100], b[10][100], c[10][100];
int i, j, k, kc, kl, il;
for i := 0 to 9 do {
  for j := 0 to 99 do { c[i][j] = 0.0; }
}
for k := 0 to 99 do {
  kc = k % 10; kl = k / 10;
  if (kc = cellid) then {
    for j := 0 to 99 do {
      brow[j] := b[kl][j];
      send(r, b[kl][j]); send(l, b[kl][j]);
    }
  }else{
    if (cellid < kc) then {
      for j := 0 to 99 do
        {recv(r, brow[j]); send(l, brow[j]);}
    }else{
      for j := 0 to 99 do
        {recv(l, brow[j]); send(r, brow[j]);}
    }
  }
  for il := 0 to 9 do {
    for j := 0 to 99 do {
      c[il][j] := c[il][j]+a[il][k]*brow[j];
    }
  }
}
```

Figure 2.3: A W2 matrix multiplication program

```
DARRAY float[100] A[100], B[100], C[100];
float brow[100];
DO(i = 0, 99){
  DO(j = 0, 99) C[i][j] = 0.0;
}
DO(k = 0, 99){
  DO(j = 0, 99) brow[j] = B[k][j];
  DO(i = 0, 99){
    DO(j = 0, 99){
      C[i][j] = C[i][j]+A[i][k]* brow[j];
    }
  }
}
```

Figure 2.4: An AL matrix multiplication program

compiler generates distributed W2 programs and manages intercell communication. Without going into detail, we can demonstrate this idea by showing a matrix multiplication program written in AL. Figure 2.4 shows an AL matrix multiplication program and Figure 2.3 shows the parallel matrix multiplication program that the AL compiler generates.

AL has three classes of data objects: scalar, array and distributed array (or DARRAY in short). Scalars and arrays are duplicated in all the cells while elements of a DARRAY are distributed among cells. Elements of a DARRAY can either be scalars or arrays. For example, the statement

```
DARRAY int A[50]
```

declares a DARRAY of integers and the statement

```
DARRAY int[50][50] B[50]
```

declares a DARRAY of 2-dimensional integer arrays. We call an element of a DARRAY a slice of the DARRAY to distinguish it from an element of a normal array. The compiler distributes slices of a DARRAY among cells of the systolic array.

AL has expression statements, compound statements, conditional statements (IF-THEN-ELSE), WHILE statements and DO-loop statements. We do not have to say much about the AL language because it is just a simple sequential programming language. Of course, we have a lot to say about the AL compiler. In the rest of this chapter, we will give an overview of the AL compiler. The following chapters will discuss each part of the AL compiler in detail.

The AL compiler generates a W2 program for each Warp cell and uses the W2 compiler to generate W1 code. The default translation of an AL program is to duplicate the computation in all the cells. Each statement of an AL program is executed by all the cells; reading a DARRAY variable is translated into broadcasting the value from the cell that has the variable to all the other cells; and writing a DARRAY variable is translated into invoking the cell that has the variable to store the value.

The compiler optimizes the program translation by transforming duplicated computation into distributed computation. In principle,

> the AL compiler assigns a statement (single or compound)
> to execute on a cell if it only references variables within
> that cell.

For example,

- If a statement only references one DARRAY slice, the compiler assigns it to execute on the cell that owns the data.

- If a DO loop references a different DARRAY slice per iteration, the compiler distributes its iterations to execute on different cells.

To illustrate program distribution, let's consider the following matrix factorization program:

```
DARRAY float[500] A[500];
float row[500];
DO(k = 0, n) {
  DO(i = k+1, n) A[k][i] = - A[k][i]/A[k][k];
  DO(i = k+1, n) row[i] = A[k][i];
  DO(j = k+1, n){
    DO(i = k+1, n){
      A[j][i] = A[j][i]+A[j][k]*row[i];
    }
  }
}
```

The statement

```
DO(i = k+1, n) A[k][i] = -A[k][i]/A[k][k];
```

only references slice $k$ of the DARRAY $A$, so the compiler assigns the statement to execute on the cell that has row $A[k]$. Execution of the statement

```
DO(i = k+1, n) row[i] = A[k][i];
```

is duplicated by the default translation and results in broadcasting the value of A[k][i] to all the cells. The **DO** loop

```
DO(j = k+1, n){
  DO(i = k+1, n){
    A[j][i] = A[j][i]+A[j][k]*row[i];
  }
}
```

references a different slice of the DARRAY $A$ every iteration. The compiler generates code for each cell to identify the slices which belong to its local space and to execute the corresponding iterations. Iterations of this DO loop are executed in parallel, because they are distributed and no communication is needed among them. Figure 2.3 shows an AL compiler generated W2 matrix factorization program for the 10-cell Warp machine. For this particular example, the compiler distributes data slices by assigning slice $A[i]$ to cell ($i$ modulo 10).

```
float a[50][500], row[500];
int k, i, fc, fi, ec, ei, lj;
for k := 0 to n do {
  if (cellid = (k % 10)) then {
    for i := k+1 to n do {
        a[k/10][i] := - a[k/10][i]/a[k/10][k]; }
  }
  for i := k+1 to n do {
    if (cellid = (k % 10)) then {
      row[i] := a[k/10][i];
      send(L, row[i]); send(R, row[i]);
    } else {
      if (cellid < (k % 10)) then {
        recv(R, row[i]); send(L, row[i]);
      }else{
        recv(L, row[i]); send(R, row[i]);
      }
    }
  }
  fc := (k+1) % 10; fi := (k+1)/10;
  if(cellid < fc) then fi := fi+1;
  ec := n % 10; ei := n / 10;
  if(cellid > ec) ei := ei-1;
  for lj := fi to ei do {
    for i := k+1 to n do {
      a[lj][i] := a[lj][i]+a[lj][k]*row[i];
    }
  }
}
```

Figure 2.5: An AL compiler generated W2 program: matrix factorization

The AL compiler derives parallelism mainly from loop distribution. To get efficient parallel code, the user has to guide the compiler with DARRAYs and DO loops. The AL compiler cannot automatically determine DARRAYs for the user. However, given a set of DARRAYs, the AL compiler automatically distributes DARRAY slices and loop iterations to exploit loop parallelism.

## 2.4   Related work

Using compilers to generate parallel code for Warp started with the Apply compiler [8]. Apply is a language for defining window operators on an image. The Apply compiler takes the definition of a window operator and generates a W2 program which reads in an image, applies the operator to all the pixels, and writes out the transformed image. Although Apply has a very limited application domain, it is an example of array-oblivious systolic array programming.

There are other research projects on compiling programs for distributed memory parallel computers. Callahan and Kennedy [4] extended FORTRAN77 with DISTRIBUTE and DECOMPOSE statements for data distribution and studied parallelizing compiler techniques for distributed memory parallel computers in general and the Intel hypercube iPSC/2 in particular. Rogers and Pingali [15] extended a functional programming language, Id Nouveau, with data distribution functions MAP, LOCAL, and ALLOC and reported a compiler for the Intel hypercube iPSC/2. Distributed data objects are also used in other parallel programming languages such as DINO [16] for the iPSC/2, and BLAZE [9] for shared memory machines such as IBM RP3 and BBN Butterfly. AL has many new ideas which are not seen in the other projects. The AL compiler uses data relations to manage distributed data objects, whereas the others require the user to explicitly define distribution functions for distributed data objects. The AL compiler has a loop distribution mechanism based on data relations, which is not seen in the other projects. The AL compiler generates code for a linear systolic array which supports fine-grain (a word per session), local (between two neighbors only), and stream-style communication, whereas the others generate code for hypercube systems which support large-grain (a message per session), global (between any two nodes in the system), and datagram-style communication.

# Chapter 3

# Data relations

In the previous chapter, we introduced the basic concepts of loop distribution. We considered the cases where each loop iteration only references one DARRAY slice. In general, we also want to distribute loops which reference multiple DARRAY slices per iteration. To do so, we have to introduce the notion of data relations first.

Given a DO loop, we say that DARRAY slices $x$ and $y$ are *related* if $x$ and $y$ are referenced in the same iteration of the DO loop. Data relation ($\sim$) is a *compatibility binary relation*; that is, it is reflexive ($x \sim x$) and symmetric ($x \sim y$ implies that $y \sim x$). However, it is not a transitive relation, that is, $x \sim y$ and $y \sim z$ does not imply that $x \sim z$.

In general, given a compatibility relation on a set D, a *compatibility class* is a subset of $D$ in which all the elements are related to each other. A *maximal compatibility class* is a compatibility class which is not a subset of any other compatibility classes. For example, consider a set of 4 elements $\{a, b, c, d\}$ and a compatibility relation $\{ a \sim b,\ b \sim c,\ c \sim d,\ d \sim a,\ b \sim d \}$. The two maximal compatibility classes derived from this compatibility relation are $\{a, b, d\}$ and $\{b, c, d\}$. Figure 3.1 shows the binary relation and maximal compatibility classes of the example where related elements are linked by edges and the two maximal compatibility classes are circled.

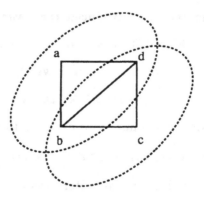

Figure 3.1: Maximal compatibility classes

As shown in this example, maximal compatibility classes are not necessarily disjoint. This is different from the equivalence classes defined by an equivalence relation, which are disjoint. This difference results from the fact that an equivalence relation is transitive (if

$x \sim y$ and $y \sim z$ then $x \sim z$) while a compatibility relation is not. For the same example, if the relation were an equivalence relation, then there would only be one equivalence class $\{a, b, c, d\}$.

Since data relation is a compatibility relation, we define the *data compatibility classes* of a DO loop to be the maximal compatibility classes of the DARRAY slices referenced in the DO loop. Intuitively, a data compatibility class corresponds to a set of DARRAY slices that are needed to execute an iteration of the DO loop. Loop distribution basically distributes data compatibility classes together with their corresponding loop iterations among cells of the systolic array. The abstractions of data relations and data compatibility classes present a simple loop distribution mechanism to the user without exposing the details of the translation. By keeping the loop distribution mechanism simple, it is possible for the user to guide the compiler to a translation that will obtain the best performance. This is very important in a systolic parallelizing compiler design because systolic array users do care about performance.

So far, we have introduced the abstractions of data relations and data compatibility classes. In the following discussion, we will describe how to use these abstractions to compile programs for a linear systolic array.

## 3.1   Linear data relations

The compiler has to select the right kind of DO loops for efficient loop
distribution because not every DO loop is suitable for distribution.
For example, consider the matrix factorization program again:

```
DARRAY float[500] A[500];
float row[500];
DO(k = 0, n){
  DO(i = k+1, n) A[k][i] = - A[k][i]/A[k][k];
  DO(i = k+1, n) row[i] = A[k][i];
  DO(j = k+1, n){
    DO(i = k+1, n){
      A[j][i] = A[j][i]+A[j][k]*row[i];
    }
  }
}
```

We do not want to distribute iterations of the outermost **DO** loop
because there is only one data compatibility class in this **DO** loop.
Obviously, loop distribution is meaningless if we only have a single
data compatibility class.

The AL compiler selectively distributes the execution of DO loops
based on data relations. For a linear systolic array, we select DO
loops with *linear data relations* for loop distribution.

Given a DO loop, if iteration $i$ of the DO loop only references DARRAY slices of the form:

$$X[d_x \cdot i + q]$$

where $d_x$ is an positive integer constant for DARRAY $X$ and $q$ is an integer constant, we say these DARRAY slices are linearly related and we call such a DO loop a DO* loop.

Notice that DARRAY slices $X[i+4]$, $X[i-3]$, and $Y[2i+1]$ are linearly related, but $X[i+2]$ and $X[2i+1]$ are not. For the same DARRAY $X$, linear data relations require $d_x$ to be a constant within the entire loop body. Also, to make the illustration clear, we introduce the symbol DO* to distinguish a distributed DO loop from a normal DO loop.

For example, consider the AL matrix multiplication program again:

```
DARRAY float[100] P[100], Q[100], R[100];
float row[100];
  DO(k = 0, 99){
    DO(j = 0, 99) row[j] = Q[k][j];
    DO*(i = 0, 99){
      DO(j = 0, 99) {
        R[i][j] = R[i][j]+P[i][k]*row[j];
      }
    }
  }
```

DARRAY slices in the DO* loop are linearly related because only $P[i]$ and $R[i]$ are referenced in iteration $i$ of the DO* loop. For another example, consider a *successive over relaxation* (SOR) [22] program:

```
DARRAY float[100] U[100];
err = 1.0; iter = 0;
while((err>zeta)&&(iter<maxiter)){
  err = 0.0; norm = 0.0;
  DO(j = 1, 98){
    DO*(i = 1, 98){
      t = U[i][j];
      U[i][j] = b*t+a*
        (U[i-1][j]+U[i][j-1]+
        U[i+1][j]+U[i][j+1]);
      norm += t**2; err += (U[i][j]-t)**2;
    }}
  err = err/norm; iter = iter + 1;
}
```

DARRAY slices in this DO* loop are linearly related because iteration $i$ of the DO* loop only references slices $U[i-1]$, $U[i]$, and $U[i+1]$.

Although linear data relations may seem to be limited in scope, they cover many of the scientific codes. To support this claim, we list all the Livermore Loop Kernels [6] which have linear data relations in Appendix A. The Livermore Loop Kernels are 24 Fortran loop kernels taken from actual production codes run at Lawrence Livermore National Laboratory. The loops represent the type of computation kernels typically found in large-scale scientific computing. Although the original Fortran programs do not declare DARRAYs, we simply determine which arrays have linear indexing patterns. Since 16 out of the 24 loop kernels have linear indexing patterns, we can at least claim that linear data relations frequently occur in real life scientific computing applications.

For a DO* loop with linearly related DARRAY slices, every iteration has its own data compatibility class, which is the union of all the DARRAY slices referenced in the iteration. For example, the data compatibility classes of the matrix multiplication DO* loop are

$$C_i = P[i] \cup R[i], \quad 0 \leq i \leq 99$$

and the data compatibility classes of SOR DO* loop are

$$C_i = U[i-1] \cup U[i] \cup U[i+1], \quad 1 \leq i \leq 98.$$

To formulate the data compatibility classes of DARRAYs $X_1, X_2, \ldots, X_m$ in a DO* loop, we have data compatibility class $C_i$ for loop iteration $i$:

$$C_i = \bigcup_{1 \leq k \leq m} \bigcup_{[l_k, q, r_k]} X_k[d_k \cdot i + q]. \tag{3.1}$$

Notice that

$$\bigcup_{[l_k, q, r_k]} X_k[d_k \cdot i + q]$$

is the union of all the DARRAY slices $X_k[d_k \cdot i + q]$ referenced in iteration $i$ of the DO* loop, where $l_k$ is the minimum value of $q$ and $r_k$ is maximum value of $q$. We use $[l_k, q, r_k]$ to denote the upper and lower bounds of $q$, it does not mean all the integers between $l_k$ and

$r_k$. The data compatibility classes in Equation 3.1 have the property that:

If

$$w \geq w_0$$

then

$$C_i \cap C_{i+w} = \emptyset$$

where

$$w_0 = \max_{1 \leq k \leq m} \lceil \frac{r_k - l_k}{d_k} \rceil$$

This property guarantees that there are no intersections among data compatibility classes whose indices are more than $w_0$ away. In Chapter 4, we will show that the intersection pattern of data compatibility classes determines the intercell communication pattern in loop distribution. Since a data compatibility class derived from linear data relations only intersects with its $w_0$ neighboring data compatibility classes, it allows us to localize the intercell communication to match the communication architecture of the linear systolic array. We will discuss intercell communication in detail in Chapter 4.

## 3.2   Joint data compatibility classes

A DO* loop defines a set of data compatibility classes, which in turn determine the distribution of DARRAY slices. If two DO* loops reference some common DARRAYs, we want to keep the distribution of those DARRAYs as static as possible because it is very expensive to change from one distribution to another at run-time. In the following, we show how to group two sets of data compatibility classes into a set of joint data compatibility classes. By statically forming a set of joint data compatibility classes at compile-time, we can avoid dynamically forming two sets of data compatibility classes at run-time.

Assume a DO* loop references DARRAYs $X_1, \ldots, X_s, Y_1, \ldots, Y_t$ and has data compatibility classes

$$
\begin{aligned}
C_i^1 \;=\; & ( \bigcup_{1 \le k \le s} \;\; \bigcup_{[xl_k, q, xr_k]} X_k[d_k \cdot i + q] \,) \; \bigcup \\
& ( \bigcup_{1 \le k \le t} \;\; \bigcup_{[yl_k, q, yr_k]} Y_k[e_k \cdot i + q] \,)
\end{aligned}
$$

and another DO* loop references DARRAYs $X_1, \ldots, X_s, Z_1, \ldots, Z_u$ and has data compatibility classes

$$
\begin{aligned}
C_i^2 \;=\; & ( \bigcup_{1 \le k \le s} \;\; \bigcup_{[xl_k', q, xr_k']} X_k[f_k \cdot i + q] \,) \; \bigcup \\
& ( \bigcup_{1 \le k \le u} \;\; \bigcup_{[zl_k, q, zr_k]} Z_k[g_k \cdot i + q] \,)
\end{aligned}
$$

Notice that DARRAYs $X_1, \ldots, X_s$ are the commonly referenced DARRAYs of the two DO* loops.

If we have the condition that

$$
\frac{d_1}{f_1} = \ldots = \frac{d_s}{f_s} = \frac{a}{b}
$$

where $gcd(a, b) = 1$, then we can group $b$ data compatibility classes from $\{C_i^1\}$ and $a$ data compatibility classes from $\{C_i^2\}$ to form a set of joint data compatibility classes $\{C_i\}$:

$$C_i = ( \bigcup_{1 \leq k \leq s} \bigcup_{[xl_k'', q, xr_k'']} X_k[b \cdot d_k \cdot i + q] ) \bigcup$$
$$( \bigcup_{1 \leq k \leq t} \bigcup_{[yl_k, q, yr_k + b \cdot e_k]} Y_k[b \cdot e_k \cdot i + q] ) \bigcup$$
$$( \bigcup_{1 \leq k \leq u} \bigcup_{[zl_k, q, zr_k + a \cdot g_k]} Z_k[a \cdot g_k \cdot i + q] )$$

where

$$xl_k'' = min(xl_k, xl_k')$$
$$xr_k'' = max(xr_k, xr_k') + b \cdot d_k$$

The joint data compatibility classes have two important properties:

1. $C_i^1 \subseteq C_{\lfloor i/b \rfloor}$ and $C_i^2 \subseteq C_{\lfloor i/a \rfloor}$,

2. they are data compatibility classes of linear data relations.

The first property guarantees that $C_{\lfloor i/b \rfloor}$ has all the needed DAR-RAY slices to execute iteration $i$ of the first DO* loop and $C_{\lfloor i/a \rfloor}$ has all the needed DARRAY slices to execute iteration $i$ of the second DO* loop. The second property guarantees that the joint data compatibility classes retain the nice properties of linear data relations, which the compiler knows how to deal with in a linear systolic array.

For example, consider the following two DO* loops:

```
DARRAY int A[100], B[200], C[50], D[50]
DO*(i = 0, 48) {
  A[2*i] = B[4*i+3]+C[i+1];
}
DO*(i = 0, 99) {
  A[i] = B[2*i]+D[i]
}
```

The data compatibility classes of the first DO* loop are

$$C_i^1 = A[2i] \cup B[4i+3] \cup C[i+1], \quad 0 \le i \le 48$$

and the data compatibility classes of the second DO* loop are

$$C_i^2 = A[i] \cup B[2i] \cup D[i], \quad 0 \le i \le 99$$

Since we have

$$\left(\frac{d_A}{f_A} = \frac{2}{1}\right) = \left(\frac{d_B}{f_B} = \frac{4}{2}\right)$$

we can group the two sets of the data compatibility classes into a set of joint data compatibility classes:

$$\begin{aligned}
C_i \quad = \quad & A[2i] \cup A[2i+1] \cup \\
& B[4i] \cup B[4i+2] \cup B[4i+3] \cup \\
& C[i+1] \cup D[2i] \cup D[2i+1]
\end{aligned}$$

where $0 \le i \le 48$.

However, consider the following two DO* loops:

```
DARRAY int A[100], B[100], C[100], D[100]
DO*(i = 0, n) {
  A[2*i] = B[4*i+3]+C[i+1];
}
DO*(i = 1, m) {
  A[i] = B[3*i]+B[3*i+1]+D[i]
}
```

We cannot use the grouping method to construct a set of joint data compatibility classes and to retain the property of linear data relations because

$$(\frac{d_A}{f_A} = \frac{2}{1}) \neq (\frac{d_B}{f_B} = \frac{4}{3}).$$

For such a case, we have to dynamically distribute some DAR-RAYs for each DO* loop, which will be discussed in the next section.

## 3.3   Scope of data compatibility classes

The default scope of a set of data compatibility classes is the DO* loop which defines them. However, by constructing joint data compatibility classes, we may extend its scope to the entire program.

If a program only has a single set or multiple unrelated sets of data compatibility classes, the scope of all its data compatibility classes is the entire program. For example, consider the matrix multiplication program:

```
DARRAY float[100] P[100], Q[100], R[100];
float row[100];
  DO(k = 0, 99){
    DO(j = 0, 99) row[j] = Q[k][j];
    DO*(i= 0, 99){
      DO(j= 0, 99){
        R[i][j] = R[i][j]+P[i][k]*row[j];
      }
    }
  }
```

Since this program only has one DO* loop, the scope of its data
compatibility classes is the entire program. Similarly, the matrix
factorization program and the SOR program in the previous sections
also have globally scoped data compatibility classes. For another
example, consider the following program:

```
DARRAY float P[200], Q[100];
  DO*(i = 1, 98){
    Q[i] = P[2*i]+0.5*(P[2*i+1]+P[2*i-1]);
  }
  DO*(i = 1, 98){
    P[2*i] = Q[i]+0.5*(Q[i+1]+Q[i-1]);
  }
```

Although there are two DO* loops in this program, the scope of their
data compatibility classes is the entire program, because the compiler
constructs a set of joint data compatibility classes from the two DO*
loops. For another example, consider the following program:

```
DARRAY float P[100], Q[100];
float a = 0.0;
  DO*(i = 1, 99){ a += P[i]; }
  DO*(i = 1, 99){
    Q[i] = Q[i]+a*(Q[i+1]+Q[i-1]);
  }
```

In this program, because the two DO* loops do not share common DARRAYs, the scope of their data compatibility classes is the entire program.

Globally scoped data compatibility classes are formed at compile-time and fixed for the entire program, thus there is no run-time overhead associated with them. On the other hand, there is some run-time overhead associated with locally scoped data compatibility classes. For example, consider the following program:

```
DARRAY int A[100], B[100], C[100], D[100]
DO* (i = 0, n) {
  A[2*i] = B[4*i+3] + C[i+1];
}
DO* (i = 1, m) {
  A[i] = B[3*i] + B[3*i+1] + D[i]
}
```

Since the compiler cannot construct joint data compatibility classes from the two DO* loops, each DO* loop defines the scope of its own set of data compatibility classes. The program starts with the data compatibility classes formed for the first DO* loop, then moves slices of DARRAYs *A* and *B* to form data compatibility classes for the second DO* loop after the first DO* loop is executed. Therefore, there are run-time communication overheads associated with the locally scoped DO* loops.

## 3.4   Summary

The abstractions of data relations and data compatibility classes present a simple loop distribution mechanism to the user without exposing the details of the translation. Intuitively, a data compatibility class corresponds to a set of DARRAY slices that are needed to execute an iteration of the DO loop. Loop distribution basically distributes data compatibility classes together with their corresponding loop iterations among cells of the systolic array. We introduced the DO* loop, a DO loop with linear data relations, for efficient loop distribution on a linear systolic array. We also introduced joint data compatibility classes and globally scoped data compatibility classes for minimizing the overhead of forming data compatibility classes at run-time. In the next chapter, we will describe all the details of loop distribution.

# Chapter 4

# Loop Distribution

The goal of loop distribution is to exploit loop parallelism. In many cases, distributed loop iterations are naturally parallel because they work on different sets of data which are totally disjoint. In these cases, there is no need for intercell communication. However, in other cases, two iterations may work on the same data objects which are duplicated in more than one cell. In these cases, intercell communication is needed to keep the the duplicated variables consistent. On a linear systolic array, it is very difficult, if not impossible, to exploit loop parallelism if random global communication is needed. Therefore, the first objective in loop distribution is to localize intercell communication. Only after obtaining local communication can we optimize the translation by load balancing and communication scheduling.

This chapter describes loop distribution techniques for a linear systolic array. In Section 4.1, we analyze the needs of intercell communication. In Section 4.2, we describe the basic loop distribution scheme for a linear systolic array. In Section 4.3, we analyze the basic forms of parallelism for distributed loops. In Section 4.4, we describe optimization techniques of load balancing and communication scheduling. In Section 4.5, we compare our loop distribution techniques with the loop concurrentization techniques for shared memory parallel computers.

## 4.1   Intercell communication

As long as we distribute loop iterations to be executed on different cells, intercell communication may be needed to keep duplicated variables consistent. In this section, we analyze the needs of intercell communication without considering the architectural limitations of a linear systolic array. We assume that the systolic array has as many cells as data compatibility classes in the DO* loop so that we can simply assign data compatibility class $C_i$ to reside on cell $i$ and loop iteration $i$ to execute on cell $i$. We assume cells are fully interconnected so that any cell can communicate with any other cell directly. We call such a loop distribution model the simple loop distribution model.

In the simple loop distribution model, each cell has a copy of the normal variables and a single data compatibility class. The compiler translates a DO* loop:

```
DO*( i = 0, n) {
    iteration i;
}
```

to a code fragment for cell $p$ ($0 \leq p \leq n$):

```
∀i < p, receive Wᵢ ∩ Cₚ from cell i;
Receive normal variables which are modified before
iteration p from cell p − 1;
Perform iteration p;
∀i ≠ p, send Wₚ ∩ Cᵢ to cell i;
Send normal variables which are modified from
iteration 0 to p to cell p + 1;
∀i > p, receive Wᵢ ∩ Cᵢ from cell i;
Globally update modified normal variables;
```

where $W_i$ is the set of variables which may be written within iteration $i$ and may be read after iteration $i$. The translation maintains the semantics of the program by keeping duplicated variables consistent after each loop iteration. For example, consider the following DO* loop:

```
DARRAY float A[100], B[100];
float sum,cs,sn;
sum = 0.0;
DO*(i = 0, 99){
    float temp;
    temp = sn*A[i]+cs*B[i];
    B[i] = temp*temp;
    sum = sum+temp;
}
```

In this example, cell $i$ has normal variables sn, cs, and sum and a data compatibility class $C_i = \{$ A[i], B[i] $\}$. Because A[i] and B[i] are not duplicated and variable temp is an iteration-local variable, we do not worry about their consistency. Although normal variables cs and sn are duplicated in all the cells, we do not update their values because they are not modified in the DO* loop. Only the normal variable sum has to be updated as the execution proceeds. As a result, for such a DO* loop, we can generate a translation for cell $p$ as shown below, where the cell variables Ap and Bp are used for DARRAY slices $A[p]$ and $B[p]$ respectively.

```
float Ap, Bp;
float sum,cs,sn;
sum = 0.0;
if(p != 0){receive sum from cell p-1; }
{
  float temp;
  temp = sn*Ap+cs*Bp;
  Bp = temp*temp;
  sum = sum+temp;
}
if(p != 99){send sum to cell p+1;}
broadcast sum from cell n to the other cells.
```

The compiler can apply conventional data flow analysis to approximate the $W$ sets. Most of the useful data flow analysis algorithms can be found in the compiler book by Aho, Sethi, and Ullman [1]. They are thus omitted in this presentation.

From the simple loop distribution model, we know that the inter-cell communication pattern of a distributed loop is solely determined by the intersection pattern of its data compatibility classes. Therefore, we can use the intersection pattern of data compatibility classes to design loop distribution schemes to match the communication architecture of the linear systolic array.

## 4.2 The basic loop distribution scheme

Consider the data compatibility classes of linearly related DARRAYs $X_1, X_2, \ldots, X_m$

$$C_i = \bigcup_{1 \leq k \leq m} \bigcup_{[l_k, q, r_k]} X_k[d_k \cdot i + q]$$

We want to distribute these data compatibility classes in a pattern that limits communication to neighboring cells only. Remember that data compatibility classes derived from linear data relations have the property that

$$C_i \cap C_{i+w} = \emptyset$$

as long as

$$w \geq w_0$$

where $w_0 = \max_{1 \leq k \leq m} \lceil (r_k - l_k)/d_k \rceil$. Based on this property, we can pack $w$ data compatibility classes into a cell to avoid duplicating

DARRAY slices among nonadjacent cells. For example, consider the
DO* loop in the SOR program again:

```
DARRAY float[500] U[500];
DO*(i = 1, 498){
  U[i][j] = a*U[i][j]+b*
       (U[i-1][j]+U[i][j-1]
       +U[i+1][j]+U[i][j+1]);
}
```

The data compatibility classes of this DO* loop are:

$$C_i = U[i - 1] \cup U[i] \cup U[i + 1], \ 1 \le i \le 498.$$

If we pack $w$ ($w \ge 2$) data compatibility classes into a cell, a DAR-
RAY slice will not appear in two nonadjacent cells. Figure 4.1 shows
that just packing two data compatibility classes into a cell will avoid
duplicating DARRAY slices in nonadjacent cells.

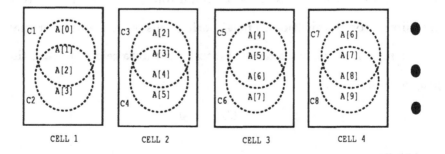

Figure 4.1: Blocking

We call such a loop distribution scheme the *blocking* scheme. For-
mally, the blocking scheme assigns data compatibility class $C_i$ to cell

$\lfloor i/w \rfloor$ of the systolic array, where $w$ is a parameter to be determined by the compiler. The blocking scheme not only localizes the communication for updating distributed variables but also localizes the communication for updating normal variables because two consecutive loop iterations are either executed in the same cell, which requires no communication, or in two neighboring cells, which only requires local communication. The parameter $w$ is determined by both the need to localize intercell communication and to accommodate all the DARRAY slices into a fixed number of cells. In most cases, there are more data compatibility classes than cells in a systolic array. If the systolic array has $P$ cells and DARRAY $X_k$ has $S_k$ data slices, in order to accommodate the entire DARRAY into the machine, we must have

$$d_k \times (w \times P) \geq S_k$$

that is,

$$w \geq \frac{S_k}{d_k P}.$$

As a result, the compiler determines the value of parameter $w$ using the following equation (for a $P$-cell linear array):

$$w = \max_{1 \leq k \leq m} \max\{\lceil \frac{r_k - l_k}{d_k} \rceil, \lceil \frac{S_k}{d_k P} \rceil\}$$

The blocking scheme also reduces the storage overhead. If a DARRAY slice belongs to two data compatibility classes which are assigned to the same cell, we only have to allocate one copy of the DARRAY slice in that cell. With the blocking scheme, only $(r_k - l_k)$ slices of the DARRAY $X_k$ are duplicated per cell and a total of $P \times (r_k - l_k)$ slices are duplicated in a $P$-cell system. For example, the SOR program has 500 DARRAY slices; we only have to duplicate 20 DARRAY slices in a 10-cell Warp machine, that is only 4% of the useful memory space.

## 4.3   Distributed loop parallelism

Loop distribution exploits two basic forms of parallelism: intraloop parallelism and interloop parallelism. To simplify the illustration, we will use the simple loop distribution model in this section. In the simple loop distribution model, each loop iteration is assigned to execute on a different cell.

### 4.3.1   Intraloop parallelism

The most important form of intraloop parallelism is DOALL parallelism where all iterations of a DO* loop are executed in parallel without communicating with each other. Given a DO* loop, if there are no intersections among its data compatibility classes and the DO* loop does not modify normal variables, then there is no need for communication and it is a parallel DOALL loop. For example, consider the DO* loop:

```
DO*(j = kp1, n){
   int i; float t;
   t = A[j][1]; A[j][1] = A[j][k]; A[j][k] = t;
   DO(i = kp1, n){A[j][i] = A[j][i]+t*col[i];}
}
```

Because $\forall j_1 \neq j_2$, $C_{j_1} \cap C_{j_2} = \emptyset$, and the DO* loop does not modify normal variables, it is a DOALL loop. The translated code for cell $j$ is shown below, where the local array Aj is used for DARRAY slice $A[j]$.

```
float Aj[500];

if (kp1 <= p <= n) then {
   int i; float t;
   t = Aj[1]; Aj[1] = Aj[k]; Aj[k] = t;
   DO(i= kp1, n){ Aj[i] = Aj[i]+t*col[i];}
}
```

Notice that there are no SEND and RECV operations in the translation and thus all the loop iterations are executed in parallel.

## 4.3.2 Interloop parallelism

If a DO* loop does not modify normal variables, a cell only has to communicate with a few neighboring cells. In this case, cells can continue their computation after the DO* loop without waiting for the completion of the entire DO* loop. This results in multiple instances of the same DO* loop or multiple unrelated DO* loops executing in parallel.

For example, consider a relaxation loop:

```
DARRAY float[100] B[100];
DO(j = 1, 98){
  DO*(i = 1, 98){
    B[i][j] = B[i][j]+0.25*
        (B[i][j-1]+B[i][j+1]+
        B[i+1][j]+B[i-1][j]);
  }
}
```

The compatibility classes and $W$ sets of this program are:

$$C_i = B[i] \cup B[i-1] \cup B[i+1], \quad 0 \le i \le 98;$$

$$W_i = \{B[i][j]\}, \quad 0 \le i \le 98.$$

Since $C_i$ only intersects with $C_{i-1}$ and $C_{i+1}$ and the DO* loop does not modify normal variables, cell $i$ only has to exchanges updates with cell $i-1$ and $i+1$. The translation of this program for cell $i$ ($1 \le i \le 98$) is listed below. In this cell program, local array Bi[0] is used for $B[i-1]$, Bi[1] for $B[i]$, and Bi[2] for $B[i+1]$, respectively.

```
float Bi[3][100];
DO (j = 1, 98){
  if (i != 1) {RECV(L, Bi[0][j]);}
  Bi[1][j] = Bi[1][j]+0.25*
      (Bi[0][j]+Bi[2][j]+
      Bi[1][j+1]+Bi[1][j-1]);
  if (i != 1) {SEND(L, Bi[1][j]);}
  if (i != 98){SEND(R, Bi[1][j]);}
  if (i != 98){RECV(R, Bi[2][j]);}
}
```

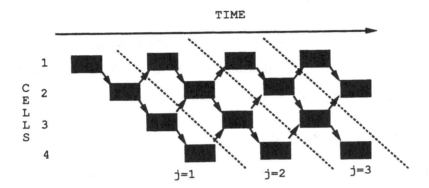

Figure 4.2: Interloop parallelism

In this translation, each cell only communicates with its two neighboring cells. The execution of an instance of the DO* loop remains sequential because cell $i$ waits for the update from cell $i-1$ before performing iteration $i$. However, after cell $i$ receives the update from cell $i+1$, it continues to execute the next iteration of the outer loop without waiting for the completion of the entire DO* loop. As a result, execution of the outer loop is overlapped. Figure 4.2 shows the parallel execution diagram of the nested loops. We use the horizontal axis for time and the vertical axis for the cell's execution. A dark box in the diagram means the cell is active in computation, otherwise it is blocked by RECV operations. An arrow in the diagram stands for communication between cells.

## 4.4   Optimization

In this section, we describe two optimization techniques for fine tuning the performance of distributed DO* loops. They are load balancing and communication scheduling.

### 4.4.1   Load balancing

Although the basic loop distribution scheme, the blocking scheme, accommodates all the DARRAY slices and localizes the intercell communication, it is not satisfactory in some cases. For example, consider the DO* loop in an LU decomposition program:

```
DARRAY float[500] A[500];
DO(k = 0, 499){
    ...
  DO* (j = k+1, 499){
    int i; float t;
    t = A[j][l]; A[j][l] = A[j][k]; A[j][k] = t;
    DO(i = k+1, n){
      A[j][i] = A[j][i]+t*col[i];
    }
  }
    ...
}
```

A data compatibility class of this DO* loop is a single slice of the DARRAY A. Using the blocking scheme, we assign 50 slices of the DARRAY $A$ to each cell of the 10-cell Warp array where cell 0 has slices 0 to 49, cell 1 has slices 50 to 99 and so on. The DO* loop invokes computation on slice $(k+1)$ to 499 of the DARRAY $A$. Since the starting index $(k + 1)$ increases with the outer loop iterations, cell 0 is only involved in the first 49 iterations of the outer loop $(k = 0$ to 48) and remains idle for the rest of the computation while the last cell is busy throughout the program. Cells are not effectively used in such a case.

The problem of unbalanced load can be improved if we do not have to block loop iterations to localize intercell communication. For example, the distributed iterations of a DOALL loop do not communicate with each other and there is no need to localize the communication. In such a case, we can use the *interleaving scheme* to balance the load. The interleaving scheme assigns data compatibility class $C_i$ to cell $(i \bmod P)$ and balances the computation load for unit-stepping DO* loops. Since most DO loops are unit-stepping, the interleaving scheme works well in most cases.

## 4.4.2   Communication scheduling

In the default translation of loop distribution, a cell receives updates from previous iterations before executing the assigned iterations. However, a cell can defer the RECV operations to avoid early synchronization if its assigned iterations do not depend on the results of previous iterations.

For example, consider the following DO* loops:

```
DARRAY float A[100], B[100];
  DO*(i = 1, 98){ A[i] = B[i-1]*B[i+1]; }
  DO*(i = 1, 98){ B[i] = A[i-1]+A[i+1]; }
```

To simplify the illustration, we assume the program fragment is mapped to 98 cells (indexed from 1 to 98) and the translation for cell $i$ ($1 \leq i \leq 98$) is:

```
float Bi[3], Ai[3]
  {
     if (i != 1){ RECV(L, Ai[0]); }
     Ai[1] = Bi[0]*Bi[2];
     if (i != 1){ SEND(L, Ai[1]); }
     if (i != 98){ SEND(R, Ai[1]); }
     if (i != 98){ RECV(R, Ai[2]); }
  }
  {
     if (i != 1){ RECV(L, Bi[0]); }
     Bi[1] = Ai[0]*Ai[2];
     if (i != 1){ SEND(L, Bi[1]); }
     if (i != 98){ SEND(R, Bi[1]); }
     if (i != 98){ RECV(R, Bi[2]); }
  }
```

Execution of each distributed loop is sequential because cell $i$ waits for the update from cell $i-1$ before it starts the computation. However, since the update is not needed for executing iteration $i$, the compiler moves the first RECV operation after the computation to parallelize the loop execution. The translation after communication scheduling is shown below:

```
float Bi[3], Ai[3]
  {
    Ai[1] = Bi[0]*Bi[2];
    if (i != 1){ SEND(L, Ai[1]);}
    if (i != 98){ SEND(R, Ai[1]);}
    if (i != 98){ RECV(R, Ai[2]);}
  }
  if (i != 1){ RECV(L, Ai[0]);}
  {
    Bi[1] = Ai[0]*Ai[2];
    if (i != 1){ SEND(L, Bi[1]);}
    if (i != 98){ SEND(R, Bi[1]);}
    if (i != 98){ RECV(R, Bi[2]);}
  }
  if (i != 1){ RECV(L, Bi[0]);}
```

Figure 4.3 shows the parallel execution diagrams of both of the translations before and after communication scheduling. Before the scheduling, although executions of the two DO* loops are pipelined, execution of each individual loop remains sequential. Whereas after the scheduling, execution of each DO* loop is parallelized. As shown in the figure, two parallel loops yield a shorter execution time than two pipelined loops.

For another example, consider the case where we modify the DO* loop in the SOR program to the following:

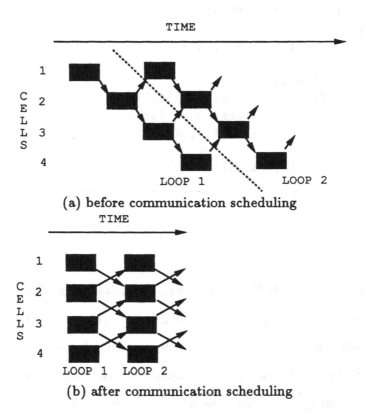

(a) before communication scheduling

(b) after communication scheduling

Figure 4.3: Communication scheduling

```
DARRAY float[100] B[100];
DO(j = 1, 98){
  DO(c = 1, 2){
    DO*(i = c, 98, 2){
      B[i][j] = B[i][j]+0.25*
               (B[i][j-1]+B[i][j+1]+
               B[i+1][j]+B[i-1][j]);
    }
  }
}
```

Since the DO* loop increases its loop index $i$ by 2, there are no dependence relations among iterations of this DO* loop. Thus, the compiler schedules all the communication operations after the computation to parallelize the loop execution. A compiler-generated W2 program for a 10-cell Warp machine is shown in Figure 4.4. In this translation, 10 data compatibility classes are assigned to each cell by the blocking scheme and part of the program is devoted to index mapping and handling communication in the boundary cells. The important thing to know in Figure 4.4 is that all the SEND and RECV operations are scheduled after the computation. Figure 4.5 shows the parallel execution diagram of this example, where iterations of the DO* loop are parallelized but not pipelined as in the original SOR program.

In general, given linear data relations and the condition that the DO* loop does not modify normal variables, it is easy to detect loop dependence to schedule the order of communication and computa-

```
float b[12][100];
g_0_bgn := cellid * 10; g_0_end := g_0_bgn + 9;
for j := 1 to 98 do {
 for c := 1 to 2 do {
  tmp_from := c; tmp_to := 98;
  cell_bgn := g_0_bgn; cell_end := g_0_end;
  cmp_bgn := cell_bgn; cmp_end := cell_end;
  tmp_i := cell_bgn - tmp_from;  tmp_q := fix(float(tmp_i)/2.0);
  tmp_mq := tmp_q * 2;  tmp_r := tmp_i - tmp_mq;
  if (tmp_r < 0) then {
   tmp_r:=tmp_r+2; tmp_q := tmp_q - 1; tmp_mq := tmp_mq - 2;}
  if (tmp_r > 0) then {cmp_bgn := cell_bgn + 2 - tmp_r;}
  tmp_i := cell_bgn - 1;
  if (tmp_i >= tmp_from && tmp_i <= tmp_to) then
   {fcpl_0_0 := 1 else fcpl_0_0 := 0;}
  tmp_i := cell_end + 0;
  if (tmp_i >= tmp_from && tmp_i <= tmp_to)then
   {fsdr_0_0 := 1 else fsdr_0_0 := 0;}
  if (cell_bgn >= tmp_from && cell_bgn <= tmp_to) then
   {fsdl_0_0 := 1 else fsdl_0_0 := 0;}
  tmp_i := cell_end + 1;
  if (tmp_i >= tmp_from && tmp_i <= tmp_to) then
   {fcpr_0_0 := 1;} else {fcpr_0_0 := 0;}
  if (cmp_bgn  > tmp_from) then
   {l_0_bgn:=cmp_bgn-cell_bgn;} else {l_0_bgn:=tmp_from-cell_bgn;}
  if (cmp_end < tmp_to) then
   {l_0_end:=cmp_end-cell_bgn;} else {l_0_end:=tmp_to-cell_bgn;}
  for l_idx := l_0_bgn to l_0_end by 2 do{
   b[l_idx+1][j] := b[l_idx+1][j] + 0.25 *
                    (b[l_idx][j]+b[l_idx+1][j-1]+
                     b[l_idx+2][j]+b[l_idx+1][j+1]);
  }
  if (fsdr_0_0 = 1) then {SEND(r, b[10][j]);}
  if (fsdl_0_0 = 1) then {SEND(l, b[1][j]);}
  if (fcpl_0_0 = 1) then {RECV(l, b[0][j]);}
  if (fcpr_0_0 = 1) then {RECV(r, b[11][j]);}
 }
}
```

Figure 4.4: An AL compiler generated W2 program: a relaxation loop

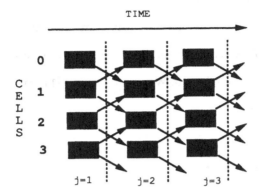

Figure 4.5: Communication scheduling: a relaxation loop

tion. The only possible data dependence of such a DO* loop has the form that:

```
DO* (i = b, e, s){
    A[d · i + q₁] = ...
        ...
    ... = A[d · i + q₂]
}
```

If there are dependence relations among iterations of this DO* loop, then there must exist two integers $x$ and $y$ that:

$$\begin{cases} d \cdot (b + s \cdot x) + q_1 = d \cdot (b + s \cdot y) + q_2 \\ x < y \end{cases}$$

Therefore, the loop iterations are dependent if and only if $(q_1 - q_2)/(d \cdot s)$ is a positive integer.

Our communication scheduling is based on the dependence relations among iterations of a DO* loop, but not among statements

within iterations. A finer grain resolution of dependence relations
would allow the compiler to schedule communication at the instruction level for fine-grain parallelism. However, fine-grain scheduling
should be and is done by the W2 cell compiler [11] and is thus omitted
in the AL compiler.

## 4.5   Related work

Although systolic arrays and shared memory multiprocessors are two
extremes in the spectrum of parallel computer architectures, loop
distribution in systolic arrays achieves the same effect as loop concurrentization in shared memory multiprocessors [21]. Of course, the
method of achieving the same effect cannot be the same for both architectures. Loop distribution relies on data compatibility relations
to determine data distribution and intercell communication, whereas
loop concurrentization relies on data dependence relations to determine inter-thread synchronization. Loop distribution assigns loop
iterations to cells at compile-time whereas loop concurrentization either pre-schedules loop iterations among processors at compile-time
or lets processors self-schedule loop iterations at run-time. Loop distribution is achieved by selecting loop iterations from a default duplicated computation within each cell, whereas loop concurrentization
is achieved by dynamically forking out parallel threads from a sin-

gle process. Although loop distribution in systolic arrays cannot be as general as loop concurrentization in shared memory multiprocessors, shared memory multiprocessors cannot scale as well as systolic arrays.

# Chapter 5

# Implementation

In this chapter, we describe features of the AL programming language which are either needed to make it practical for use or are introduced to simplify the compiler implementation. In Section 5.1, we describe how an AL program interfaces with the outside world. In section 5.2, we describe the implementation of loop distribution. In section 5.3, we introduce the ALIGN* statement for efficient block data transfer. In section 5.4, we introduce the accumulative operators for parallel accumulation. Finally, we describe a debugging tool for AL program development.

## 5.1 External interface

In the Warp programming environment, the Warp machine is a computation server and AL is a language for defining its services. An AL program is defined with input and output parameters and is invoked

from a client program using a remote procedure call mechanism. For
example, the following AL program:

```
program madd (A, B) (C)
DARRAY float[100] A[100], B[100], C[100];
{
  body of the AL program
}
```

has two input parameters, A and B, and one output parameter, C.
In an AL program invocation, the input parameters are first copied
from the host and distributed to the Warp cells; then the AL program
is executed; and finally the output parameter is copied back to the
client. The following is a simple client program which calls the AL
program to execute on the Warp machine.

```
#include<stdio.h>
#include<math.h>
#include<monitor.h>
static float A[100][100], B[100][100];
static float C[100][100];
main(){
int i, j;
  for(i = 0; i <= 99; i++){
    for(j = 0; j <= 99; j++){
      A[i][j] = (float)(random()%1000);
      B[i][j] = (float)(random()%1000);
    }}
  wu_call(''madd'', A, B, C);
}
```

## 5.2 Compiling DO* loops

To simplify data flow analysis, AL asks the user to explicitly declare iteration-local variables in a DO* loop. Iteration-local variables are allocated at the beginning of each iteration and deallocated at the end of each iteration and thus do not need to be kept consistent between loop iterations. After filtering out iteration-local variables, we further simplify the compiler design by limiting loop distribution to those DO loops which do not modify normal variables or only modify normal variables by accumulators. For the other cases, the user can either restructure the program to get around this restriction or simply let the loop run sequentially. We will discuss accumulators in Section 5.4.

Since normal variables are not written in the DO* loop, the data flow analyzer only has to consider how DARRAYs are written in the DO* loop to approximate the $W$ sets. The data flow analyzer implemented can precisely calculate the written area on a dimension of a DARRAY if

- its index is invariant in the DO* loop, or

- its index is a linear function of a DO loop index whose bounds are invariant in the DO* loop.

Otherwise the compiler conservatively assumes the entire dimension is modified. For example, the data flow analyzer knows that the following DO* loop modifies $B[i][1 : n : 2]$ in iteration $i$.

```
DARRAY float[500] B[500];
DO*(i = 1, n, 2){
  int j;
  DO(j = 1, n, 2){
    B[i][j] = B[i][j]+0.25*
        (B[i-1][j]+B[i][j-1]+B[i+1][j]+B[i][j+1]);
  }
}
```

However, it assumes the following DO* loop modifies the entire DARRAY slice $B[i][*]$ in iteration $i$:

```
DARRAY float[500] B[500];
int IA[500];
DO*(i = 1, n, 2){
  DO(j = 1, n, 2){
    B[i][IA[j]] = B[i][j]+0.25*
        (B[i-1][j]+B[i][j-1]+B[i+1][j]+B[i][j+1]);
  }
}
```

## 5.3   The ALIGN* statement

To improve the performance of global communication, the ALIGN* statement is introduced for copying a block of data between a DARRAY and a normal array or between two DARRAYs. An ALIGN* statement:

```
ALIGN*(idx = from, to, step) lhs = rhs;
```

is exactly the same as the DO statement:

$$DO(idx = from, to, step) \; lhs = rhs;$$

However, the compiler will try to generate a block data transfer for the ALIGN* statement instead of the default translation which transfers a single data item per loop iteration.

For example, the following is a matrix multiplication program $(C = C + A \times B)$ written in AL:

```
mm(A, B, C, n)(C){
DARRAY float[100] A[100], B[100];
DARRAY float[100] C[100];
float brow[100];
int i, k;
DO(k = 0, n){
  ALIGN*(j = 0, n) brow[j] = B[k][j];
  DO*(i = 0, n){
    int j;
    DO(j = 0, n){
      C[i][j] = C[i][j]+A[i][k]*brow[j];
    }
  }
}
}
```

where the ALIGN* statement is used instead of a DO statement for copying a row of the $B$ matrix to the normal array *brow*.

## 5.4   Parallel accumulation

Although the AL compiler does not distribute DO* loops which modify normal variables, it does distribute those DO* loops which employ
normal variables as accumulators. For example, consider the following DO* loop:

```
DARRAY float A[500], B[500];
float dot;
    .
DO*(i = 0, 499){
   dot = dot+A[i]*B[i];
}
    .
```

We assume that the variable dot is defined somewhere else and will
be used after the DO* loop. Execution of the DO* loop is sequential
because each cell has to wait for the newly updated value of dot
before starting its own execution. However, if we assume that floating point addition is commutative $(a + b = b + a)$ and associative[1]
$((a + b) + c = a + (b + c))$, cells can compute the partial sums in parallel and then combine all the partial sums together for the final result. In such a case, no immediate global updates of the variable dot
are needed until the end of the DO* loop.

In AL, four accumulative operators: +=, *=, max=, min= are
defined:

---

[1]Floating point addition is not associative in computer arithmetic.

- *lhs* += *rhs* means *lhs* = *lhs* + *rhs*;

- *lhs* *= *rhs* means *lhs* = *lhs* * *rhs*;

- *lhs* max= *rhs* means *lhs* = max(*lhs*, *rhs*);

- *lhs* min= *rhs* means *lhs* = min(*lhs*, *rhs*).

Under the assumption of commutativity and associativity, accumulation can be executed in an arbitrary order. If the possible round off error in floating point addition is not an issue, the previous example can be rewritten using an accumulator:

```
DARRAY float A[500], B[500];
float dot;
       .
DO*(i = 0, 499){
   dot += A[i]*B[i];
}
```

Parallel accumulation of this example is executed in 3 steps:

1. Initialization: All but the first cell resets its local copy of the variable dot to the base value 0.0.

2. Local accumulation: Each cell executes the iterations belonging to its local data space and adds values to its local copy of the variable dot.

3. Global combination: Partial results are combined together from two ends of the systolic array toward the center of the systolic array. The final sum then propagates bidirectionally from the central cell back to all the other cells.

If the variable under accumulation is an array object, the AL data flow analyzer tries to identify the modified area in the array and minimize the overhead of initialization and global combination. However the compiler can not always determine the precise area by analysis. For example, consider a histogram program:

```
DARRAY int[256] img[256];
int hist[256];
int i;
DO(i = 10, 20) hist[i] = 0;
DO*(i = 0, 255){
  int j;
  DO(j = 0, 255) hist[img[i][j]]+=1;
}
```

The compiler can never know the run-time values of the array img. For the best performance, AL allows the user to define the area of accumulation explicitly through the INIT* statement for initialization, and the COMBINE* statement for combining. As shown below, one can use the INIT* and COMBINE* statements to tell the compiler that the values of the array img in the histogram example are between 10 to 20 so that the compiler does not have to initialize and combine 256 elements of the entire hist array:

```
DARRAY int[256] img[256];
int hist[256];
int i;
INIT*{for(i = 10, 20) hist[i] = 0;}
DO*(i = 0, 255){
  int j;
  DO(j = 0, 255) hist[img[i][j]]+=1;
}
COMBINE*{for(i = 10, 20) hist[i];}
```

## 5.5 Program debugging

We have implemented a simple translator which treats DARRAYS as normal arrays and literally translates an AL program to a C program. The user can debug the AL program by debugging its equivalent C program on a sequential machine. Although the user may not be able to find all the bugs from the equivalent C program, he or she probably can find most of the bugs without using the Warp machine. This simple idea is extremely useful in AL programming. Our experience of converting the LINPACK SVD routine into an AL program demonstrates this fact. A few hours were spent debugging the C program on a workstation. After that, the AL program was compiled into a 1000-line W2 program. We received the correct result from the same set of test data in the first run on the Warp machine. In comparison, two weeks are normally needed to debug a 100-line hand-written W2 program.

# Chapter 6

# Evaluation

The AL compiler has been successfully used to compile many scientific programs to the Warp machine [19, 20]. In this chapter, we present some examples and evaluate the achieved performance. These samples include representative algorithms for matrix computations (LU decomposition, QR decomposition and singular value decomposition), signal processing (2D-FFT) and numerical solutions of partial differential equations (SOR).

The matrix computation routines are taken from LINPACK and the other programs are taken from textbooks. (All the Fortran programs were first manually translated into the AL syntax.) For each program, we conducted the following experiments:

1. compiled the existing program to run on a single Warp cell;

2. measured the single-cell execution time;

3. selected DARRAYs to produce an AL program for multiple cells;

4. compiled the modified programs to run on multiple (2 to 10) cells;

5. measured the multi-cell execution times;

6. calculated the speedup factors from the measured execution times.

An AL program for multiple cells may not be exactly the same as the one for a single cell because the multi-cell version may have extra data movement for loop distribution. However, data movement does not change the computational procedure. Both the single-cell and multi-cell versions use the same computational procedure and produce the same results. Because the AL compiler generates W2 code which is then compiled by the W2 compiler, it is the W2 compiler and the Warp cell architecture, not the AL compiler, which determine the absolute performance of a program. As a result, we evaluate the AL compiler by the speedup factors, not by the absolute performance (MFLOPS). To make an honest speedup measurement, we use the original code, not the modified code, to define the single-cell execution time because the modified code may have extra data copying which is not needed in the original program.

To analyze the experimental results, we divide the single-cell execution time $T(1)$ into 2 parts:

$$T(1) = D + S$$

Parameter D is the time spent on the DO* loops. Parameter S is the time spent on the rest of the computation. The compiler parallelizes the DO* loops but duplicates the rest of the computation in all the cells. As a result, the execution time of the $p$-cell parallel program $T(p)$ cannot be shorter than $S + (D/p)$, that is,

$$T(p) \geq S + \frac{D}{p}$$

This inequality is generally applicable to all the AL programs and the parameters $S$ and $D$ can be measured experimentally from the single cell execution. The value of $T(p)$ can get close to the lower bound $S + D/p$ if (1) the compiler can parallelize all the DO* loops and (2) the overhead of communication is relatively small. While most of the DO* loops in the sample programs can be parallelized, the overhead of communication is not small enough to be neglected. Because the extra global communication such as normal variable updates and global combining cannot be overlapped with the computation, we can add the total overhead of global communication $C(p)$ to the right hand side of the inequality:

$$T(p) \geq S + \frac{D}{p} + C(p)$$

The value of $C(p)$ cannot be directly measured, unlike the other two parameters $S$ and $D$. Experiments were designed using a case by case basis to measure the value of $C(p)$. The lower bound on the $p$-cell execution time results in an upper bound on the best expected speedup of using $p$ cells:

$$Speedup(p) \leq \frac{S + D}{S + \frac{D}{p} + C(p)} \tag{6.1}$$

Problem size affects the speedup figures because the parameters $S$, $D$ and $C(p)$ are functions of the problem size. The 32K words of data memory in a Warp cell can at most accommodate a square matrix of size $180 \times 180$. Although the limited data memory size makes it impossible to study the effect of problem size, small matrices are good for the purpose of performance evaluation because the overheads of the duplicated computation and global communication are visible. Big matrices tend to diminish the relative importance of the overhead in parallel execution and generate linear speedup curves which do not tell anything interesting.

## 6.1 Matrix computations

The implementation of three LINPACK routines is presented in this section. They are the LU decomposition (SGEFA), QR decomposition (SQRDC) and singular value decomposition (SSVDC) routines.

For these routines, we decompose the working matrix into columns and use ALIGN* to move columns of the working matrix to normal variables. In this section, if we use a two dimensional array A to represent a matrix, then $A[i][*]$ stands for a column of the matrix.

### 6.1.1 LU decomposition

An LU decomposition program for a single cell is listed in Appendix B.1.1 and the same program for multiple cells is listed in Appendix B.1.2. In the multi-cell program, the matrix $A$ is decomposed into columns and the ALIGN* statement is used to create the DO* loop.

The tables of execution time and the speedup curves of the LU decomposition program are shown in Figure 6.1 for matrices of size $100 \times 100$ and in Figure 6.2 for matrices of size $180 \times 180$. The parenthesized values in the first row of the tables are the values of the parameter $S$; that is, the execution time of the program in Appendix B.1.1 without the boxed DO loop. The parenthesized values in the other rows are the values of $C(p)$. The values of $C(p)$ are measured by executing the following program.

| cells | time(msec) |
|-------|------------|
| 1 (S) | 411 (31) |
| 2 (C) | 228 (5) |
| 3 (C) | 166 (6) |
| 4 (C) | 131 (5) |
| 5 (C) | 114 (6) |
| 6 (C) | 101 (6) |
| 7 (C) | 92 (6) |
| 8 (C) | 84 (5) |
| 9 (C) | 80 (6) |
| 10 (C) | 75 (6) |

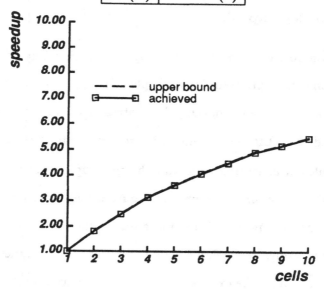

Figure 6.1: LU decomposition: 100 × 100 matrix

| cells | time(msec) |
|:---:|:---:|
| 1 (S) | 2199 (91) |
| 2 (C) | 1160 (15) |
| 3 (C) | 815 (17) |
| 4 (C) | 635 (15) |
| 5 (C) | 532 (17) |
| 6 (C) | 462 (17) |
| 7 (C) | 413 (17) |
| 8 (C) | 372 (15) |
| 9 (C) | 345 (17) |
| 10 (C) | 322 (17) |

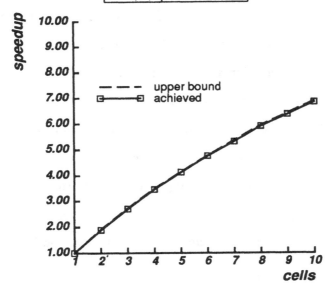

Figure 6.2: LU decomposition: $180 \times 180$ matrix

```
DO(k = 0, nm1){
   ALIGN*(i = k, n) col[i] = a[k][i];
}
```

The global communication overheads of using 2, 4 and 8 cells are less than those of the others. This is because an index transformation ($k$ modulo $p$) is used to determine which cell has column $k$ of the matrix $a$. If $p = 2$, 4, or 8, the modulo operation uses 3 integer operations and takes 3 cycles. In the other cases, the modulo operation uses floating point operations and takes 54 cycles. Other than the overhead of index transformation, the time spent on communication is almost the same for 2 cells to 10 cells. This is because the ALIGN* statement is translated to a global data broadcast.

Although the compiler is able to generate parallel programs achieving speedups close to the expected upper bound defined by the inequality 6.1, the LU decomposition program spends a significant amount of time on sequential computation. For the case of a $100 \times 100$ matrix, the cost of sequential computation limits the best possible speedup of using 10 cells to less than 6. The time spent on sequential computation is mainly determined by the following two loops:

- the pivoting loop:

```
DO(i = kp1, n){
   if(abs(col[i])>smax)
     {smax = abs(col[i]); l = i;}
}
```

- and the scaling loop:

```
DO(i = kp1, n)
  {col[i] = t*col[i];}
```

The results of scheduling these two loops are analyzed using the W2 profiler. The profiler counts the number of cycles used by an iteration of a DO loop. In this case, the results are 24 cycles for the pivoting loop and 3 cycles for the scaling loop. The pivoting loop is slow because its execution can not be pipelined.

The pivoting loop can be parallelized. The pivoting loop searches for the index of the maximal absolute value in the array *col*. Because the array *col* is duplicated in all the cells, we can assign each cell to do $\frac{1}{p}th$ (assume a $p$-cell system) of the searching, and then combine the local maxima to determine the global maximum. In general, if a parallel DO loop only references normal variables, there are two options in compilation. One option is to duplicate the computation in all the cells. The other option is to distribute the iterations among cells, and then update the modified normal variables at the end. The first option saves the overhead of communication but ignores the parallelism. The second option parallelizes the computation but risks the potentially high overhead in global communication. The compiler should take the second option for the pivoting loop because only two variables are modified while $24 \times (n - k)$ cycles are needed by duplication. However, the AL compiler takes the first option for all cases and leaves the second option for future research. For a quick solution, we borrowed the ideas used in the data parallel programming languages and defined a function for parallel pivoting. The improved results are shown in Figure 6.3 and 6.4 where the global communication overhead includes both the ALIGN* statement and the final combining step of the parallel pivoting. The experimental results show that parallel pivoting reduces the cost of the sequential computation and thus improves the overall performance.

| cells | time(msec) |
|:-----:|:----------:|
| 1 (S) | 411 (5) |
| 2 (C) | 219 (6) |
| 3 (C) | 153 (8) |
| 4 (C) | 118 (7) |
| 5 (C) | 99 (8) |
| 6 (C) | 86 (9) |
| 7 (C) | 77 (9) |
| 8 (C) | 68 (8) |
| 9 (C) | 64 (9) |
| 10 (C) | 60 (10) |

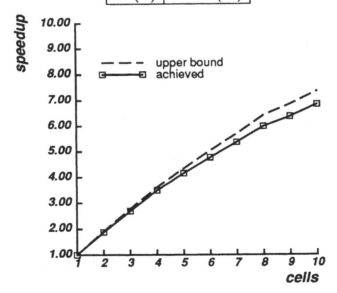

Figure 6.3: LU decomposition with parallel pivoting: $100 \times 100$ matrix

| cells | time(msec) |
|---|---|
| 1 (S) | 2199 (14) |
| 2 (C) | 1130 (17) |
| 3 (C) | 771 (20) |
| 4 (C) | 588 (19) |
| 5 (C) | 481 (21) |
| 6 (C) | 409 (22) |
| 7 (C) | 358 (23) |
| 8 (C) | 318 (21) |
| 9 (C) | 290 (24) |
| 10 (C) | 266 (24) |

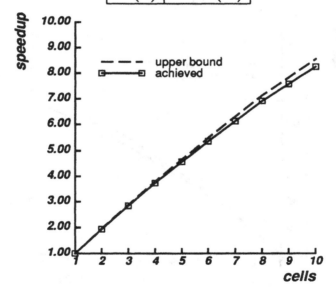

Figure 6.4: LU decomposition with parallel pivoting: 180 × 180 matrix

## 6.1.2  QR decomposition

A QR decomposition program for a single cell is listed in Appendix B.2.1 and the same program for multiple cells is listed in Appendix B.2.2. To make the presentation easy to understand, the AL programs presented here are simplified from the LINPACK SQRDC routine, although the algorithm remains complete. The original LINPACK program was written for general rectangular matrices and has an option for pivoting. The AL programs are the same routine for square matrices using the option of not pivoting. As in the LU decomposition program, the matrix $A$ is decomposed into columns. The first ALIGN* statement moves a column of the matrix A to a normal array *col*. The first DO* loop does not reference any DARRAYs but directs the compiler to parallelize the accumulation. The second DO* loop is directly derived from the single-cell program. The second ALIGN* statement moves the normal array *col* back to the matrix $A$.

The tables of execution time and speedup curves of the QR decomposition program for matrices of size $100 \times 100$ and $180 \times 180$ are shown in Figure 6.5 and 6.6. The value of $C(p)$ is measured by executing the following program:

```
DO(l = 0, n){
  ALIGN*(i = 1, n) col[i] = a[l][i];
  COMBINE*{ nrmxl;}
}
```

In the QR decomposition, the overhead of sequential computation and global communication is relatively small compared to the total execution time, and good speedups are expected. Again, the compiler is able to generate speedups close to the expected upper bound.

| cells | time (msec) |
|:-----:|:-----------:|
| 1 (S) | 733 (5) |
| 2 (C) | 375 (6) |
| 3 (C) | 258 (7) |
| 4 (C) | 198 (6) |
| 5 (C) | 164 (8) |
| 6 (C) | 141 (8) |
| 7 (C) | 124 (8) |
| 8 (C) | 111 (7) |
| 9 (C) | 102 (8) |
| 10 (C) | 94 (9) |

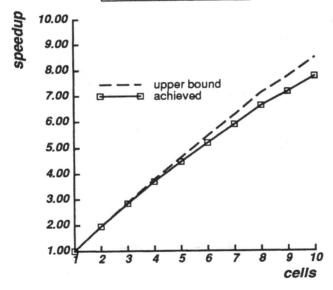

Figure 6.5: QR decomposition: $100 \times 100$ matrix

| cells | time (msec) |
|:-----:|:-----------:|
| 1 (S) | 3663 (14) |
| 2 (C) | 1839 (16) |
| 3 (C) | 1246 (19) |
| 4 (C) | 946 (17) |
| 5 (C) | 770 (19) |
| 6 (C) | 651 (20) |
| 7 (C) | 567 (20) |
| 8 (C) | 502 (18) |
| 9 (C) | 454 (21) |
| 10 (C) | 414 (21) |

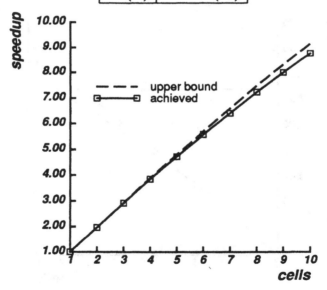

Figure 6.6: QR decomposition: 180 × 180 matrix

## 6.1.3 Singular value decomposition

Only the first part of the LINPACK SVD program is used for this evaluation. The second part of the algorithm is inherently sequential and is not interesting for the purpose of performance evaluation. The first part of the SVD program transforms a full matrix into bidiagonal form. An SVD matrix bidiagonization program for a single cell is listed in Appendix B.3.1 and the same program for multiple cells is listed in Appendix B.3.2. The matrix $A$ is decomposed into columns. The first ALIGN* statement moves a column of the matrix A to the normal array *col* and the second ALIGN* statement moves a row of the matrix $A$ to the normal array *row*. There are several DO* loops in the program and some of them use accumulators.

The tables of execution time and speedup curves of the SVD bidiagonalization program for matrices of size $100 \times 100$ and $175 \times 175$ are shown in Figure 6.7 and 6.8. The values of $C(p)$ are measured by executing the following program.

```
DO(1 = 0, n-1){
   ALIGN*(i = 1, n) col[i] = a[l][i];
   COMBINE* {nrmxl;}
   ALIGN*(i = l+1, n) row[i] = a[i][l];
   COMBINE* {nrmxl;}
   COMBINE*{for(i = l+1, n) work[i];}
}
```

For this program, the time spent on sequential computation is much less than the overhead of global communication. However, both overheads are relatively small compared to the overall execution time and good speedups are expected. Again, the compiler is able to generate high performance parallel programs close to the expected upper bound.

| cells | time (msec) |
|-------|-------------|
| 1 (S) | 1539 (10) |
| 2 (C) | 818 (26) |
| 3 (C) | 568 (28) |
| 4 (C) | 440 (27) |
| 5 (C) | 367 (30) |
| 6 (C) | 318 (31) |
| 7 (C) | 283 (32) |
| 8 (C) | 255 (30) |
| 9 (C) | 237 (34) |
| 10 (C) | 220 (35) |

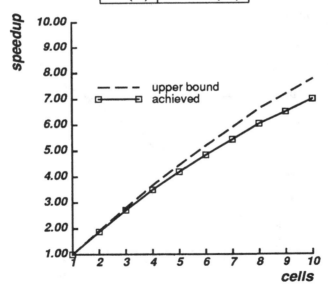

Figure 6.7: SVD bidiagonalization: $100 \times 100$ matrix

| cells | time (msec) |
|---|---|
| 1 (S) | 7372 (26) |
| 2 (C) | 3812 (72) |
| 3 (C) | 2597 (76) |
| 4 (C) | 1985 (73) |
| 5 (C) | 1624 (78) |
| 6 (C) | 1382 (80) |
| 7 (C) | 1209 (82) |
| 8 (C) | 1077 (80) |
| 9 (C) | 981 (85) |
| 10 (C) | 902 (87) |

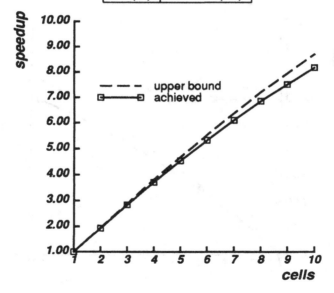

Figure 6.8: SVD bidiagonalization: 175 × 175 matrix

## 6.2   2D Fast Fourier Transform

A 2D FFT program for a single cell is listed in Appendix B.4.1. and
the same program for multiple cells is listed in Appendix B.4.2. This
program is for a complex image of size 64 × 64 where the arrays
*gr* (real part) and *gi* (imaginary part) together define the complex
image. Notice that a 64 × 64 is the largest complex image which
is $2^n \times 2^n$ in size and can fit in the 32K-word memory of a Warp
cell; a 128 × 128 complex image needs 32K words but leaves no space
for other variables. The 2D FFT program first performs 1D FFT
on each row of the image then performs 1D FFT on each column of
the image. In the multiple cell program, we transpose the matrix
between the two phases.

The execution time and speedup curve of the 2D FFT programs
are shown in Figure 6.9. In this example, the cost of duplicated
computation is negligible but the overhead of communication is of the
same order as the total execution time. The cost of communication
is the time spent on the matrix transposition.

```
DO(1 = 0, 63){
  ALIGN* (j = 0, 63) gr[j][1] = wr[1][j];
  ALIGN* (j = 0, 63) gi[j][1] = wi[1][j];
}
```

Whereas a linear array is not good for transposing an image from
column distribution to row distribution, the compiler parallelizes the

| cell    | Time(ms)    |
|---------|-------------|
| 1 (S)   | 46.4 (0.0)  |
| 2 (C)   | 31.8 (8.5)  |
| 3 (C)   | 23.4 (7.4)  |
| 4 (C)   | 17.5 (5.9)  |
| 5 (C)   | 15.4 (6.0)  |
| 6 (C)   | 13.6 (5.6)  |
| 7 (C)   | 12.7 (5.3)  |
| 8 (C)   | 10.4 (4.6)  |
| 9 (C)   | 10.9 (5.0)  |
| 10 (C)  | 10.0 (4.9)  |

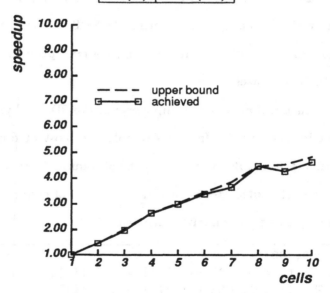

Figure 6.9: 2D FFT, 64x64 complex image

rest of the computation and achieves speedup close to the expected upper bound. The performance when using 2, 4 and 8 cells is closer to the upper bound than in other cases because 64 columns or rows can be evenly distributed among 2, 4 or 8 cells but not in the other cases.

## 6.3   Partial differential equation solvers

Three relaxation methods for solving a model partial differential
equation problem are presented in this section. These methods are
representative in the solution of large PDE problems and are frequently used to benchmark the performance of supercomputers.

The problem is to solve the partial differential equation:

$$\frac{\partial^2 U(x,y)}{\partial^2 x} + \frac{\partial^2 U(x,y)}{\partial^2 y} = 0$$

on a square region with known boundary values. The square region
is descretized into a mesh of 180 × 180 grid points (the maximum
mesh which a Warp cell can accommodate). The value at each mesh
point is an unknown except on the boundaries and there are 31,682
unknowns.

### 6.3.1   SOR

One method for solving the PDE problem is the successive over relaxation (SOR) [22] method. The SOR algorithm for this problem
is shown in Appendix B.5.1. The time for executing 200 relaxation
steps and the speedup curve are shown in Figure 6.10.

In this example, although all the loop iterations are distributed
among cells, the cost of duplicated computation cannot be neglected
because of the overhead of starting up the inner most loop. The

overhead includes moving data from memory to registers, computing the starting address of array references and starting up the loop pipeline. In the previous examples, all the DO* loops have at least one inner loop in their body and the overheads of starting up inner most loops are distributed and parallelized. In this example, the DO* loop itself is the inner most loop and the overhead of starting up the loop is thus duplicated. As a result, the cost of duplicated computation cannot be neglected. On the other hand, the overhead of global communication in this example is very small.

The compiler pipelines the outer DO loop and generates fine grain communication (a word at a time) between neighboring cells. The achieved speedup differs from the predicted upper bound because the overhead of fine grain local communication is not included in the upper bound formula. Although the overhead of local communication cannot be measured experimentally; it is clear that as the number of cells increases, the granularity of computation decreases and the overhead of local communication becomes relatively significant.

## 6.3.2   Line SOR

A modified relaxation algorithm for solving the same problem is the line relaxation algorithm. The line relaxation algorithm divides the rows of the mesh into odd and even sets such that all the even rows or odd rows can be processed in parallel. The algorithm is listed in

| cell | Time(sec) |
|------|-----------|
| 1 (S) | 42.75 (1.30) |
| 2 (C) | 22.80 (.005) |
| 3 (C) | 16.22 (.005) |
| 4 (C) | 12.72 (.005) |
| 5 (C) | 10.62 (.005) |
| 6 (C) | 9.22 (.005) |
| 7 (C) | 8.29 (.005) |
| 8 (C) | 7.60 (.005) |
| 9 (C) | 6.90 (.005) |
| 10 (C) | 6.44 (.005) |

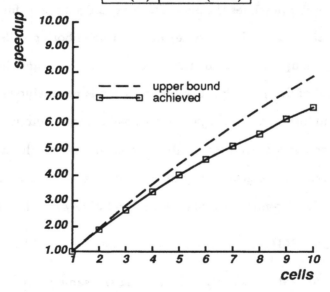

Figure 6.10: SOR

Appendix B.5.2. The time for executing 200 relaxation steps and the speedup curve are shown in Figure 6.11.

In the line SOR algorithm, both the cost of duplicated computation and that of global communication are negligible because the DO* loop is not an inner most loop. Together with the fact that iterations of the DO* loop are not dependent, an ideal linear speedup is expected. The experimental results show that the compiler generates parallel programs close to the expected performance. One interesting point in the speedup curve is the performance of 8 cells. The achieved performance of 8 cells is a little bit worse than the others because of the unbalanced load. Matrix $U$ has 180 rows which can be evenly partitioned among 2, 3, 4, 5, 6, 9 or 10 cells. For 7 cells, the first 6 cells process 26 ($\lceil 180/7 \rceil = 26$) rows each and the last cell processes 24 ($180 = 26x6+24$) rows. In this case, the effect of unbalanced load is not so significant. For 8 cells, the first 7 cells process 23 ($\lceil 180/8 \rceil = 23$) rows each and the last cell processes 19 ($180 = 23x7+19$) rows. Here, the effect of the unbalanced load becomes significant.

## 6.3.3 Two-color SOR

Although the line SOR method achieves close to linear speedups, its absolute performance is poor because the sequential inner most loop cannot take advantage of the pipelined functional units in the

| cell | Time(sec) |
|------|-----------|
| 1    | 42.25     |
| 2    | 21.43     |
| 3    | 14.35     |
| 4    | 11.02     |
| 5    | 8.66      |
| 6    | 7.24      |
| 7    | 6.29      |
| 8    | 5.81      |
| 9    | 4.87      |
| 10   | 4.40      |

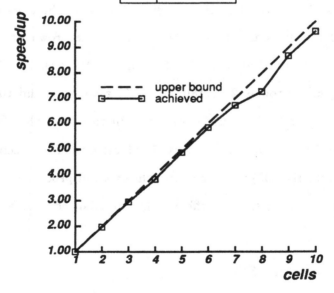

Figure 6.11: Line SOR

Warp cell. One simple way to improve the absolute performance is to use the two-color relaxation method as shown in the AL program of Appendix B.5.3. For this program, the AL compiler distributes and parallelizes the outer loop among cells, and the W2 compiler pipelines the inner loop inside each cell. As shown in Figure 6.12, the execution time is significantly improved while the linear speedup is retained.

| cell | Time(sec) |
|------|-----------|
| 1    | 12.67     |
| 2    | 6.45      |
| 3    | 4.34      |
| 4    | 3.34      |
| 5    | 2.64      |
| 6    | 2.21      |
| 7    | 1.92      |
| 8    | 1.77      |
| 9    | 1.50      |
| 10   | 1.36      |

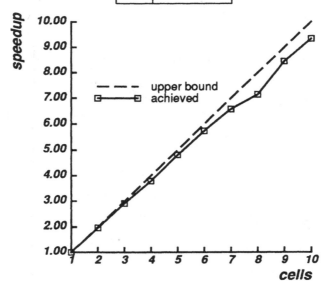

Figure 6.12: Two-color SOR

## 6.4 Summary

We evaluated the AL compiler by measuring the achieved speedups for a sample of scientific programs. We chose these programs not only because they are representative of scientific computing but also because they are good examples for illustrating the overhead in the generated parallel code. These sample programs are matrix computations (LU decomposition, QR decomposition and singular value decomposition), 2D-FFT, and relaxation algorithms for solving partial differential equations. For each program, we measured the execution time and the overhead of sequential computation and global communication in the Warp machine. We used these experimental results to analyze the speedup curve for each program. The results show that the compiler can generate efficient parallel code close to the best achievable.

# Chapter 7

# Conclusions

The AL compiler is a powerful programming tool for the Warp systolic array because it manages all the details of systolic array programming for the user. The user programs the entire systolic array as if it were a sequential computer and the compiler manages the details of program decomposition, data distribution, intercell communication, process synchronization, and load balancing.

In AL, the user guides the compiler with DARRAYs and DO loops to exploit distributed loop parallelism. The AL compiler cannot automatically determine DARRAYs for the user. However, given a set of DARRAYs, the AL compiler automatically distributes DARRAY slices and loop iterations to exploit loop parallelism.

The abstractions of data relations and data compatibility classes present a simple loop distribution mechanism to the user without exposing the details of the translation. Intuitively, a data compatibility class corresponds to a set of DARRAY slices that are needed to execute an iteration of the DO loop. Loop distribution basically distributes data compatibility classes together with their corresponding loop iterations among cells of the systolic array. By keeping the loop distribution mechanism simple, it is possible for the user to guide the compiler to a translation that will obtain the best performance.

The user sees a simple loop distribution mechanism, the compiler manages all the details of the program translation. The compiler localizes intercell communication, balances the workload, and optimizes the schedule of communication. The AL compiler exploits both intra and inter loop parallelism with loop distribution.

We take a high-level, array-oblivious approach to systolic array programming. However, we do not sacrifice the goal of pursuing performance. By a sample of representative algorithms in scientific computing, we have shown that the AL compiler can generate efficient parallel W2 codes which are almost identical to what the user would have written by hand. For example, the AL compiler generates parallel code for the LINPACK LU decomposition (SGEFA) and QR decomposition (SQRDC) routines with a nearly 8-fold speedup on the 10-cell Warp array for matrices of size 180 × 180.

AL only begins a research area in systolic array parallelizing compilers. Many interesting problems remain to be solved. We have not considered multi-dimensional DARRAYs, nested DO* loops and procedure calls in the current implementation. Multi-dimensional DARRAY and nested DO* loops may be extended from the current implementation. Procedure calls require a meaningful parameter passing scheme. Although we cannot use the call-by-reference scheme for distributed objects, we can use the call-by-value-result scheme and apply compiler optimization to avoid unnecessary data copying.

There are many other research areas worth pursuing in systolic array paralleling compilers. AL only covers a very limited scope in systolic array programming. There are many other ways to use a systolic array which can not be programmed in AL. Kung said that there are five ways (computational models) to use the Warp machine [10]. AL only covers one and a half out of Kung's five. Also, the next-generation systolic arrays, for example, the iWarp machine [3], will have 2D configurations. Parallelizing compilers will play an even more important role in 2D systolic array programming because it is simply too difficult to do 2D programming using a cell programming language. I believe, as long as there are demands for systolic array computing, there are demands for powerful programming tools such as parallelizing compilers to simplify the task of systolic array programming.

# Bibliography

[1] A. V. Aho, R. Sethi, and J. D. Ullman. *Compilers: Principles, Techniques and Tools*. Addison Wesley, Reading, MA, 1985.

[2] M. Annaratone, E. Arnould, T. Gross, H. T. Kung, M. Lam, O. Menzilcioglu, and J. A. Webb. The Warp Computer: Architecture, Implementation and Performance. *IEEE Transactions on Computers*, C-36(12):1523–1538, December 1987.

[3] S. Borkar, R. Cohn, G. Cox, S. Gleason, T. Gross, H. T. Kung, M. Lam, B. Moore, C. Peterson, J. Pieper, L. Rankin, P. S. Tseng, J. Sutton, J. Urbanski, and J. Webb. iWarp: An Integrated Solution to High-Speed Parallel Computing. In *Proceedings of Supercomputing '88*, pages 330–339, Orlando, Florida, November 1988. IEEE Computer Society and ACM SIGARCH.

[4] D. Callahan and K. Kennedy. Compiling Programs for Distributed-Memory Multiprocessors. *The Journal of Supercomputing*, 2:151–169, October 1988.

[5] J. Delosme and I. Ipsen. Design Methodology for Systolic Arrays. In *Proceedings of SPIE Symposium, Vol. 696, Advanced Algorithms and Architectures for Signal Processing*, pages 245–259, August 1986.

[6] J. Foe. An Analysis of the Computational and Parallel Complexity of Livermore Loops. *Parallel Computing*, 7:163–186, 1988.

[7] J. A. B. Fortes and D. I. Moldovan. Parallelism Detection and Transformation Techniques Useful for VLSI Algorithms. *Journal of Parallel and Distributed Computing*, 2(3):277–301, 1985.

[8] L. G. C. Hamey, J. A. Webb, and I. C. Wu. Low-level Vision on Warp and the Apply Programming Model. In *Parallel Computation and Computers for Artificial Intelligence*, pages 185–199. Kluwer Academic Publishers, 1987.

[9] C. Koelbel, P. Mehrotra, and J. V. Rosendale. Semi-Automatic Domain Decomposition in BLAZE. In *Proc. of the 1988 International Conference on Parallel Processing*, pages 521–524, August 1987.

[10] H. T. Kung. The Warp Computer. In *Biological and Artificial Intelligence Systems*, pages 319–424. ESCOM Science Publishers B.V., 1988. Edited by E. Clementi and S. Chin.

[11] M. Lam. Compiler Optimizations for Asynchronous Systolic Array Programs. In *Proc. Fifteenth Annual ACM Symposium on Principles of Programming Languages*, Jan. 1988.

[12] M. Lam and J. Mostow. A Transformational Model of VLSI Systolic Design. In *Proceedings of the 6th International Symposium on Computer Hardware Description Languages and their Applications*, pages 65–77. IFIP, May 1983.

[13] M. S. Lam. *A Systolic Array Optimizing Compiler*. PhD thesis, Carnegie Mellon University, May 1987.

[14] P. Quinton. Automatic Synthesis of Systolic Arrays from Uniform Recurrent Equations. In *Conference Proceedings of the 11th Annual International Symposium on Computer Architecture*, pages 208–214, Ann Arbor, Michigan, June 1984.

[15] A. Rogers and K. Pingali. Process Decomposition Through Locality of References. In *ACM SIGPLAN'89 Conference on Programming Language Design and Implementation.*, pages 69–80, June 1989.

[16] M. Rosing, R. Schnabel, and R. Weaver. Dino: Summary and Examples. In *The Third Conference on Hypercube Concurrent Computers and Applications*, pages 472–481, ACM, New York, 1988.

[17] P. S. Tseng. *Ph.D. Thesis Proposal: High Level Programming Constructs for Systolic Arrays.* Carnegie Mellon University, December 1987.

[18] P. S. Tseng. *AL: A Programming Tool for Warp.* Carnegie Mellon University, 1988.

[19] P. S. Tseng. Iterative Sparse Linear System Solvers on Warp. In *Proc. of the 1988 International Conference on Parallel Processing*, volume 1, pages 32–38, August 1988.

[20] P. S. Tseng, M. Lam, and H. T. Kung. The Domain Parallel Computation Model on Warp. In *Proceedings of SPIE Symposium, Vol. 977, Real-Time Signal Processing XI*, pages 130–137. Society of Photo-Optical Instrumentation Engineers, August 1988.

[21] M. Wolfe. *Optimizing Supercompilers for Supercomputers.* MIT Press, Cambridge, Massachusetts, 1989.

[22] D. Young. *Iterative Solution of Large Linear Systems.* Academic Press, New York, 1971.

# Appendix A

# Linear data relations in Livermore Loops

**Kernel 1**

```
      DARRAY REAL X(n), Y(n), Z(n+11)
      DO* 1 k = 1,n
 1        X(k)= Q+Y(k)*(R*Z(k+10)+T*Z(k+11))
```

**Kernel 3**

```
      DARRAY REAL Z(n), X(n)
      Q = 0.0
      DO* 3 k = 1,n
 3        Q = Q+Z(k)*X(k)
```

**Kernel 5**

```
      DARRAY REAL X(n), Z(n), Y(n), X(n)
      DO* 5 i = 2,n
 5        X(i) = Z(i)*(Y(i)-X(i-1))
```

**Kernel 7**

```
      DARRAY REAL X(n), U(n+5), Z(n), Y(n)
      DO* 7 k = 1,n
      X(k) = U(k)+R*(Z(k)+R*Y(k))+
      T*(U(k+3)+R*(U(k+2)+R*U(k+1))+
      T*(U(k+6)+R*(U(k+5)+R*U(k+4))))
 7    CONTINUE
```

## Kernel 8

```
DARRAY REAL DU1(n), DU2(n), DU3(n)
DARRAY REAL(3) U11(n+1), U21(n+1), U31(n+1)
DARRAY REAL(3) U12(n+1), U22(n+1), U32(n+1)
DO 8 kx = 2,3
DO* 8 ky = 2,n
 DU1(ky) = U11(kx,ky+1)-U11(kx,ky-1)
 DU2(ky) = U21(kx,ky+1)-U21(kx,ky-1)
 DU3(ky) = U31(kx,ky+1)-U31(kx,ky-1)
 U12(kx,ky) = U11(kx,ky)+A11*DU1(ky)+
   A12*DU2(ky)+A13*DU3(ky)+
   SIG*(U11(kx+1,ky)-2.*U11(kx,ky)+
        U11(kx-1,ky))
 U22(kx,ky) = U21(kx,ky)+A21*DU1(ky)+
   A22*DU2(ky)+A23*DU3(ky)+
   SIG*(U21(kx+1,ky)-2.*U21(kx,ky)+
        U21(kx-1,ky))
 U32(kx,ky) = U31(kx,ky)+A31*DU1(ky)+
   A32*DU2(ky)+A33*DU3(ky)+
   SIG*(U31(kx+1,ky)-2.*U31(kx,ky)+
        U31(kx-1,ky))
 8 CONTINUE
```

## Kernel 9

```
DARRAY REAL(13) PX(n)
DO* 9 i = 1,n
 PX(1,i) =   DM28*PX(13,i)+
   DM27*PX(12,i)+DM26*PX(11,i)+
   DM25*PX(10,i)+DM24*PX( 9,i)+
   DM23*PX( 8,i)+DM22*PX( 7,i)+
   CO*(PX(5,i)+ PX(6,i))+PX(3,i)
 9 CONTINUE
```

## Kernel 10

```
DARRAY REAL(13) PX(n)
DARRAY REAL(5)  CX(n)
DO* 10 i = 1,n
AR      =        CX(5,i)
BR      = AR-PX(5,i)
PX(5,i) = AR
CR      = BR-PX(6,i)
PX(6,i) = BR
AR      = CR-PX(7,i)
PX(7,i) = CR
```

```
      BR       = AR-PX(8,i)
      PX(8,i)  = AR
      CR       = BR-PX(9,i)
      PX(9,i)  = BR
      AR       = CR-PX(10,i)
      PX(10,i) = CR
      BR       = AR-PX(11,i)
      PX(11,i) = AR
      CR       = BR-PX(12,i)
      PX(12,i) = BR
      PX(14,i) = CR-PX(13,i)
      PX(13,i) = CR
   10 CONTINUE
```

## Kernel 11

```
      DARRAY REAL X(n), Y(n)
      X(1)= Y(1)
      DO* 11 k = 2,n
   11   X(k)= X(k-1)+Y(k)
```

## Kernel 12

```
      DARRAY REAL X(n), Y(n)
      DO* 12 k = 1,n
   12   X(k)= Y(k+1)-Y(k)
```

## Kernel 13

```
      DARRAY REAL(4) P(n)
      DO* 13 ip = 1,n
       i1 = P(1,ip)
       j1 = P(2,ip)
       i1 = 1+MOD2N(i1,64)
       j1 = 1+MOD2N(j1,64)
       P(3,ip) = P(3,ip)+B(i1,j1)
       P(4,ip) = P(4,ip)+C(i1,j1)
       P(1,ip) = P(1,ip)+P(3,ip)
       P(2,ip) = P(2,ip)+P(4,ip)
       i2 = P(1,ip)
       j2 = P(2,ip)
       i2 = MOD2N(i2,64)
       j2 = MOD2N(j2,64)
       P(1,ip) = P(1,ip)+Y(i2+32)
       P(2,ip) = P(2,ip)+Z(j2+32)
       i2 = i2+E(i2+32)
       j2 = j2+F(j2+32)
```

```
         H(i2,j2) = H(i2,j2)+1.0
13     CONTINUE
```

## Kernel 14

```
       DARRAY INTEGER IX(n),IR(n)
       DARRAY REAL XI(n), EX1(n), DEX1(n)
       DARRAY REAL XX(n), RX(n), VX(n)
       DO* 141  k = 1,n
       IX(k) = INT(GRD(k))
       XI(k) = FLOAT(IX(k))
       EX1(k) = EX(IX(k))
       DEX1(k) = DEX(IX(k))
141    CONTINUE
       DO* 142  k = 1,n
       VX(k) = VX(k)+EX1(k)
       +(XX(k)-XI(k))*DEX1(k)
       XX(k) = XX(k)+VX(k) +FLX
       IR(k) = XX(k)
       RX(k) = XX(k)-IR(k)
       IR(k) = MOD2N(IR(k),512)+1
       XX(k) = RX(k)+IR(k)
142    CONTINUE
       DO* 14 k = 1,n
       RH(IR(k)) = RH(IR(k))+1.0-RX(k)
       RH(IR(k)+1) = RH(IR(k)+1)+RX(k)
14     CONTINUE
```

## Kernel 18

```
       DARRAY REAL(JN+1)
       ZA(KN+1), ZB(KN+1), ZM(KN+1), ZP(KN+1),
       ZQ(KN+1), ZR(KN+1), ZU(KN+1), ZV(KN+1),
       ZZ(KN+1)
       DO* 70 k = 2,KN
       DO 70 j = 2,JN
       ZA(j,k) = (ZP(j-1,k+1)+ZQ(j-1,k+1)-
       ZP(j-1,k)-ZQ(j-1,k))*
       (ZR(j,k)+ZR(j-1,k))/
       (ZM(j-1,k)+ZM(j-1,k+1))
       ZB(j,k) = (ZP(j-1,k)+
       ZQ(j-1,k)-ZP(j,k)-ZQ(j,k))*
       (ZR(j,k)+ZR(j,k-1))/
       (ZM(j,k)+ZM(j-1,k))
70     CONTINUE
       DO* 72  k = 2,KN
       DO 72  j = 2,JN
       ZU(j,k) = ZU(j,k)+
       S*(ZA(j,k)*(ZZ(j,k)-ZZ(j+1,k))-
```

```
                ZA(j-1,k) *(ZZ(j,k)-ZZ(j-1,k))-
                ZB(j,k)   *(ZZ(j,k)-ZZ(j,k-1))+
                ZB(j,k+1) *(ZZ(j,k)-ZZ(j,k+1)))
            ZV(j,k) = ZV(j,k)+
                S*(ZA(j,k)*(ZR(j,k)-ZR(j+1,k))-
                ZA(j-1,k) *(ZR(j,k)-ZR(j-1,k))-
                ZB(j,k)   *(ZR(j,k)-ZR(j,k-1))+
                ZB(j,k+1) *(ZR(j,k)-ZR(j,k+1)))
72      CONTINUE
        DO* 75 k = 2,KN
        DO 75  j = 2,JN
            ZR(j,k) = ZR(j,k)+T*ZU(j,k)
            ZZ(j,k) = ZZ(j,k)+T*ZV(j,k)
75      CONTINUE
```

## Kernel 20

```
        DARRAY REAL
            Y(n), G(n), U(n), V(n), VX(n)
            X(n), XX(n), W(n), Z(n)
        DO* 20 k = 1,n
        DI = Y(k)-G(k)/( XX(k)+DK)
        DN = 0.2
        IF(DI.NE.0.0) THEN
            DN = AMAX1( 0.1,AMIN1( Z(k)/DI, 0.2))
        ENDIF
        X(k) = ((W(k)+V(k)*DN)* XX(k)+U(k))/
                (VX(k)+V(k)*DN)
        XX(k+1) = (X(k)- XX(k))*DN+ XX(k)
20      CONTINUE
```

## Kernel 21

```
        DARRAY REAL(n) PX(n), CX(n)
        DO* 21 j = 1,n
         DO 21  k = 1,n
          DO 21 i = 1,n
            PX(i,j) = PX(i,j)+VY(i,k)*CX(k,j)
        21 CONTINUE
```

## Kernel 22

```
        DARRAY REAL U(n), V(n), W(n), X(n), Y(n)
        EXPMAX = 20.0
        U(n) = 0.99*EXPMAX*V(n)
        DO* 22 k = 1,n
            Y(k) = U(k)/V(k)
            W(k) = X(k)/( EXP( Y(k)) -1.0)
22      CONTINUE
```

**Kernel 23**

```
      DARRAY REAL(n+1)
       ZA(n+1), ZB(n+1), ZR(n+1), ZU(n+1)
       ZV(n+1), ZZ(n+1)
      DO* 23  j = 2,n
      DO 23  k = 2,n
       QA = ZA(k,j+1)*ZR(k,j) +
       ZA(k,j-1)*ZB(k,j) + ZA(k+1,j)*ZU(k,j)+
       ZA(k-1,j)*ZV(k,j) +ZZ(k,j)
 23    ZA(k,j) = ZA(k,j) +.175*(QA -ZA(k,j))
```

# Appendix B

# Benchmark programs

## B.1 LU decomposition

### B.1.1 Single cell

```
sgefa(a, n)(a, ipvt, info){
  float a[180][180];
  int n,info,ipvt[180];
  int nm1,i,j,k,l,kp1;
  float smax,t,w;
    info=-1; nm1=n-1;
    DO(k=0, nm1){
      kp1=k+1 ;
      l=k; smax=abs(a[k][l]);
      /* PIVOTING */
      DO(i=kp1, n){
        if(abs(a[k][i])>smax)
          {smax=abs(a[k][i]); l=i;}
      }
      ipvt[k]=l;
      if (a[k][l] == 0.0) {
        info=k;
      }else{
        t=a[k][l];
        a[k][l]=a[k][k];
        a[k][k]=t;
        t=-1.0/ t;
        /* SCALING */
        DO(i=kp1, n){a[k][i]=t*a[k][i];}
```

```
/* ELIMINATION */
DO(j=kp1, n){
  t=a[j][1];
  a[j][1]=a[j][k];
  a[j][k]=t;
  DO(i=kp1, n)
    {a[j][i]=a[j][i]+t*a[k][i];}
}
```

```
      }
    }
    ipvt[n]=n;
    if(a[n][n] == 0.0){info=n;}
}
```

## B.1.2   Multiple cells

```
sgefa(a, n)(a, ipvt, info){
  DARRAY float[180] a[180];
  int n,info,ipvt[180];
  int nm1,i,j,k,l,kp1;
  float smax,t,col[180];
    info=-1; nm1=n-1;
    DO(k=0, nm1){
      ALIGN*(i=k, n)col[i]=a[k][i];
      kp1=k+1;
      l=k; smax=abs(col[l]);
      /* PIVOTING */
      DO(i=kp1, n){
        if(abs(col[i])>smax)
          {smax=abs(col[i]); l=i;}
      }
      ipvt[k]=l;
      if (col[l]==0.0){
        info=k;
      }else{
        t=col[l];
        col[l]=col[k];
        col[k]=t;
        t=-1.0/t;
        /* SCALING */
        DO(i=kp1, n){col[i]=t*col[i];}
```

```
/* ELIMINATION */
DO*(j=kp1,n){
  int i; float t;
  t=a[j][l];
  a[j][l]=a[j][k];
  a[j][k]=t;
  DO(i=kp1, n)
    {a[j][i]=a[j][i]+t*col[i];}
}
ALING*(i=k, n) a[k][i]=col[i];
    }
  }
ipvt[n]=n;
if(a[n][n] == 0.0){info=n;}
}
```

## B.2   QR decomposition

### B.2.1   Single cell

```
sqrdc(a, n)(a, qraux){
  float a[180][180];
  float qraux[180], nrmxl,t;
  int n,i,j,k,l;
  DO(l=0, n){
      nrmxl=0.0;
      DO(i=1, n){
        nrmxl=nrmxl+a[l][i]*a[l][i];}
    nrmxl= sqrt(nrmxl);
    if (nrmxl != 0.0){
        if (a[l][l]<0.0){nrmxl= -nrmxl;}
        t=1.0/nrmxl;
        DO(i=1, n){a[l][i]=t*a[l][i];}
        a[l][i]=1.0+a[l][i];
          DO(j=(l+1), p){
            t=0.0;
            DO(i=1, n){
              t=t+a[l][i]*a[j][i];}
            t=-t/a[l][l];
            DO(i=1, n){
              a[j][i]=a[j][i]+t*a[l][i];}
          }
        qraux[l]=a[l][l];
        a[l][l]= -nrmxl;
    }
  }
}
```

### B.2.2   Multiple cells

```
sqrdc(a, n)(a, qraux){
  DARRAY float[180] a[180];
  float qraux[180], col[180], nrmxl,t;
  int n,i,j,k,l;
  DO(l=0, n){
```

```
ALIGN*(i=1, n) col[i]=a[l][i];
┌─────────────────────────────────────────┐
│  init*{nrmxl=0.0;}                        │
│  DO*(i=1, n){                             │
│    nrmxl += col[i]*col[i];}               │
│  combine*{nrmxl;}                         │
└─────────────────────────────────────────┘
nrmxl=sqrt(nrmxl);
if (nrmxl != 0.0){
    if (col[l]<0.0){nrmxl= -nrmxl;}
    t=1.0/nrmxl;
    DO(i=1, n){col[i]=t*col[i];}
    col[l]=1.0+col[l];
    ┌───────────────────────────────────────┐
    │  DO*(j=(l+1), p){                      │
    │    float t; int i;                     │
    │    t=0.0;                              │
    │    DO(i=1, n){                         │
    │      t=t+col[i]*a[j][i];}              │
    │    t=-t/col[l];                        │
    │    DO(i=1, n){                         │
    │      a[j][i]=a[j][i]+t*col[i];}        │
    │  }                                     │
    └───────────────────────────────────────┘
    qraux[l]=col[l];
    col[l]= -nrmxl;
    ALIGN*(i=1, n)a[l][i]=col[i];
  }
 }
}
```

# B.3   Singular value decomposition

## B.3.1   Single cell

```
svd1(a, n)(a, s, e){
   float a[175][175];
   float col[175],row[175];
   float work[175],s[175],e[175];
   int n,l,ll,i,j,k,m;
   float nrmxl,t;
     DO(l = 0, (n - 1)){
          ll = l + 1;
          nrmxl = 0.0;
          DO(i = l, n)
             {nrmxl+=a[l][i]*a[l][i];}

          nrmxl= sqrt(nrmxl);
          if (nrmxl > 0.0) {
               if (a[l][l] < 0.0){nrmxl=-nrmxl;}
               t = 1.0/nrmxl;
               DO(i = l, n){a[l][i]=t*a[l][i];}
               a[l][l] = 1.0+a[l][l]);
               DO(j = ll, n){
                  t = 0.0;
                  DO(i = l, n)
                    {t=t+a[l][i]*a[j][i];}
                  t = t/col[l];
                  DO(i = l, n)
                    {a[j][i]=a[j][i]-t*a[l][i];}
               }
          }
          s[l]=- nrmxl;
          nrmxl = 0.0;
          DO(i = ll, n)
             {nrmxl += a[i][l]*a[i][l];}

          nrmxl = sqrt(nrmxl);
          if (nrmxl > 0.0) {
               if (a[ll][l] < 0.0)
                 {nrmxl=-nrmxl;}
               t = 1.0/nrmxl;
               DO(i = ll, n)
```

```
        {a[i][1]=t*a[i][1];}
      a[l1][1]=1.0+a[l1][1];
```
```
  DO(i=l1,n){work[i]=0.0;}
  DO(j=l1,n){
    DO(i = l1, n)
      {work[i]+=a[j][1]*a[j][i];}}
```
```
  DO(j = l1, n){
    t = (a[j][1]/a[l1][1]);
    DO(i = l1, n)
      {a[j][i]=a[j][i]-t*work[i];}
  }
```
```
    }
    e[1] = -nrmxl;}
    s[n] = a[n][n];
}
```

## B.3.2  Multiple cell

```
svd1(a, n)(a, s, e){
  DARRAY float[175] a[175];
  float col[175],row[175];
  float work[175],s[175],e[175];
  int n,l,l1,i,j,k,m;
  float nrmxl,t;
    DO(l = 0, (n - 1)){
      l1 = l + 1;
      ALIGN*(i = l, n) col[i] = a[l][i];
```
```
    init* {nrmxl = 0.0;}
    DO*(i = l, n)
      {nrmxl+=col[i]*col[i];}
    combine*{nrmxl;}
```
```
      nrmxl= sqrt(nrmxl);
      if (nrmxl > 0.0) {
          if (col[l] < 0.0)
            {nrmxl=-nrmxl;}
          t = 1.0/nrmxl;
          DO(i = l, n){col[i]=t*col[i];}
          col[l] = (1.0 + col[l]);
```

```
DO*(j = 11, n){
    float t; int i;
    t = 0.0;
    DO(i = 1, n)
      {t=t+col[i]*a[j][i];}
    t = t/col[1];
    DO(i = 1, n)
      {a[j][i]=a[j][i]-t*col[i];}
}
```

```
}
s[1]=- nrmx1;
ALIGN*(i = 11, n)row[i]=a[i][1];
```

```
init*{nrmx1 = 0.0;}
DO*(i = 11, n)
  {nrmx1+=row[i]*row[i];}
combine*{nrmx1;}
```

```
nrmx1 = sqrt(nrmx1);
if (nrmx1 > 0.0) {
    if (row[11] < 0.0){nrmx1=-nrmx1;}
    t = 1.0/nrmx1;
    DO(i = 11, n){row[i]=t*row[i];}
    row[11]=1.0+row[11];
```

```
    init*{for(i=11,n)work[i]=0.0;}
    DO*(j = 11, n){
        int i;
        DO(i = 11, n)
          {work[i]+=row[j]*a[j][i];}}
    combine*{for(i=11,n)work[i];}
```

```
    DO*(j = 11, n){
      float t; int i;
      t = (row[j]/row[11]);
      DO(i = 11, n)
        {a[j][i]=a[j][i]-t*work[i];}}
```

```
}
e[1] = -nrmx1;}
s[n] = a[n][n];
```

```
}
```

# B.4   2D Fast Fourier Transform

## B.4.1   Single cell

```
c2dfft(cor, coi, brev, gr, gi)
      (gr, gi)
{
float cor[6][32], coi[6][32];
int brev[64];
float br[64], bi[64];
float gr[64][64], gi[64][64];
int j,l;
```

```
    DO(l = 0, 63){
      DO(j = 0, 63)
        br[j] = gr[l][brev[j]];
        bi[j] = gi[l][brev[j]];}
      C1DFFT(br, bi, cor, coi);
      DO(j = 0, 63)
        gr[l][j] = br[j];
        gi[l][j] = bi[j];}
    }
```

```
    DO(l = 0, 63){
      DO(j = 0, 63){
        br[j] = gr[brev[j]][l];
        bi[j] = gi[brev[j]][l];}
      C1DFFT(br, bi, cor, coi);
      DO(j = 0, 63){
        gr[j][l] = br[j];
        gi[j][l] = bi[j];}
    }
```

```
}
C1DFFT(br, bi, cor, coi)
float br[64], bi[64];
float cor[6][32], coi[6][32];
{
    float b1r[64], b1i[64];
```

```
float tr, ti;
int i, j, k;
DO(k = 0, 5, 2){
  DO(j =0, 31){
    tr = br[2*j+1]*cor[k][j]-
            bi[2*j+1]*coi[k][j];
    ti = br[2*j+1]*coi[k][j]+
            bi[2*j+1]*cor[k][j];
    b1r[j] = br[2*j] + tr;
    b1i[j] = bi[2*j] + ti;
    b1r[j+32] = br[2*j] - tr;
    b1i[j+32] = bi[2*j] - ti;
    }
  DO(j =0, 31){
    tr = br[2*j+1]*cor[k][j]-
            bi[2*j+1]*coi[k][j];
    ti = br[2*j+1]*coi[k][j]+
            bi[2*j+1]*cor[k][j];
    b1r[j] = br[2*j] + tr;
    b1i[j] = bi[2*j] + ti;
    b1r[j+32] = br[2*j] - tr;
    b1i[j+32] = bi[2*j] - ti;
    }
  }
}
```

## B.4.2   Multiple cell

```
c2dfft(cor, coi, brev, gr, gi)
      (gr, gi)
{
float cor[6][32], coi[6][32];
int brev[64];
DARRAY float[64] gr[64], gi[64];
DARRAY float[64] wr[64], wi[64];
int l;
```

```
DO*(1 = 0, 63){
  float br[64], bi[64];
  int j;
  DO(j = 0, 63){
    br[j] = gr[1][brev[j]];
    bi[j] = gi[1][brev[j]];}
  C1DFFT(br, bi, cor, coi);
  DO(j = 0, 63){
    wr[1][j] = br[j];
    wi[1][j] = bi[j];}
}
```

```
DO(1 = 0, 63){
  ALIGN*(j = 0, 63)gr[j][1] = wr[1][j];
  ALIGN*(j = 0, 63)gi[j][1] = wi[1][j];
}
```

```
  DO*(1 = 0, 63){
    float br[64], bi[64];
    int j;
    DO(j = 0, 63){
      br[j] = gr[1][brev[j]];
      bi[j] = gi[1][brev[j]];}
    C1DFFT(br, bi, cor, coi);
    DO(j = 0, 63){
      gr[1][j] = br[j];
      gi[1][j] = bi[j];}
  }
}
```

# B.5 Partial differential equation solvers

## B.5.1 SOR

```
SOR(U, maxiter, zeta)(U, err,iter){
  DARRAY float[180] U[180];
  int i,j,k,maxiter,iter;
  float err,norm,t,zeta;
  err = 1.0; iter = 0;
```

```
while((err>zeta)&&(iter<maxiter)){
  init*{err = 0.0; norm = 0.0;}
  DO(j = 1, 178){
    DO*(i = 1, 178){
      float t;
      t = U[i][j];
      U[i][j]= 0.49*(U[i-1][j]+
          U[i][j-1]+U[i+1][j]+U[i][j+1])
          -0.97*t;
      norm+=t**2; err+=(U[i][j]-t)**2;
    }
  }
  combine*{err; norm;}
  err = err/norm; iter = iter + 1;
}}
```

## B.5.2   Line SOR

```
LSOR(U, maxiter, zeta)(U, err,iter){
  DARRAY float[180] U[180];
  int i,k,,maxiter,iter;
  float err,norm,t,zeta;
  err = 1.0; iter = 0;
  while((err>zeta)&&(iter<maxiter)){
    init*{err = 0.0; norm = 0.0;}
    DO(k=1,2){
      DO*(i= k,178,2){
        float t; int j;
        DO(j=1, n){
          t = U[i][j];
          U[i][j]= 0.49*(U[i-1][j]+
            U[i][j-1]+U[i+1][j]+U[i][j+1])
            -0.97*t;
          norm+=t**2;
          err+=(U[i][j]-t)**2;
        }
      }
    }
```

```
   combine*{err; norm;}
   err = err/norm; iter = iter + 1;
}}
```

## B.5.3   Two-color SOR

```
SOR2C(U,maxiter,zeta)(U,err,iter){
   DARRAY float[180] U[180];
   int i,k,maxiter,iter;
   float err,norm,t,zeta;
   err = 1.0; iter = 0;
   while((err>zeta)&&(iter<maxiter)){
      init*{err = 0.0; norm = 0.0;}
      DO(k=1,2){
         DO*(i= k,178,2){
            float t; int l, j;
            DO(l = 1, 2){
               DO(j=1,178,2){
                  t=U[i][j];
                  U[i][j]=0.49*(U[i-1][j]+
                     U[i][j-1]+U[i+1][j]+U[i][j+1])
                     -0.97*t;
                  norm += t**2;
                  err+=(U[i][j]-t)**2;
            }}
         }
      }
      combine*{err; norm;}
      err = err/norm; iter = iter + 1;
}}
```

# Index